Christian College Library
Columbia, Missouri

Wildlife Communities

CLARENCE J. HYLANDER

Wildlife Communities

from the Tundra to the Tropics in North America

Illustrated with photographs and maps, drawings, and diagrams by the author

HOUGHTON MIFFLIN COMPANY BOSTON / 1966

Other Books by the Author

The World of Plant Life
The Macmillan Wildflower Book
Trees and Trails
Flowers of Field and Forest
Sea and Shore
Animals in Armor
Insects on Parade
Animals in Fur
Feathers and Flight
Fishes and Their Ways

COPYRIGHT © 1966 BY DORIS D. HYLANDER
ALL RIGHTS RESERVED INCLUDING THE RIGHT
TO REPRODUCE THIS BOOK OR PARTS THEREOF IN ANY FORM
LIBRARY OF CONGRESS CATALOG CARD NUMBER: 66–19940
PRINTED IN THE U.S.A.

CONTENTS

PREFACE: A New Look at Nature	1
PART I WILDLIFE AND ITS ENVIRONMENT	7
1 An Ecological Adventure	8
2 An Inventory of American Wildlife	14
3 Adaptation of Life to Its Physical Environment	35
4 Adaptation of Life to Its Living Environment	53
5 The Biotic Community	70
6 The Biotic Community in Action	77
PART II THE BIOMES OF NORTH AMERICA	93
7 Life and Its Physical Environment in North America	94
8 The Tundra	113
9 The Conifer Forest	132
10 The Deciduous Forest	168
11 At the Edge of the Tropics	212
12 The Grasslands	241
13 The Desert	263
14 Wildlife Sanctuaries in North America	296
Index	335

Wildlife Communities

Our wildlife heritage includes many protected wilderness communities such as this, which reaches from conifer forest through alpine tundra to snow-capped mountain summits. The Sierran subalpine forest, Mount Adams, Cascade Mountains, Washington. (*Credit: U.S. Forest Service*)

Preface

A New Look at Nature

This is the story of the role of environment in determining the appearance, living habits, and distribution of the plants and animals that make up our wildlife heritage. Why do certain kinds of animals live in one part of the country and not another? How are plants adapted to live in such different environments as the tundra and the tropics? How are the many different species of plants and animals adjusted to living with each other? Most important of all, how has man played a part as a member of the environment, affecting the abundance and survival of the living things which share the land with us?

Much light is shed on these and other questions of a similar nature by the biological science of *ecology*, which is concerned with the relations of organisms to their environment. Environment, the sum of all the conditions surrounding an organism, is more than a physical one of air, land, and water. For every organism, the presence of other forms of life is also a vital part of the environment. Ecology therefore deals with both the nonliving and the living surroundings, and the many interrelations among plants and animals and man. The ecological view gives

Environment is the sum of all the surrounding conditions that affect an organism.

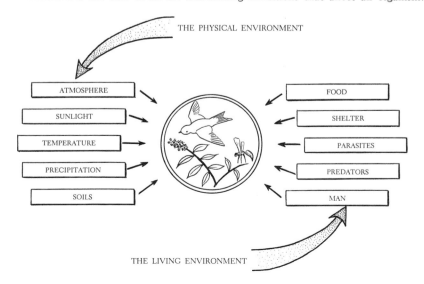

an insight which enables us better to appreciate the problems facing wildlife and to evaluate human attempts to change it.

Many centuries ago, when biology was natural history, the individual organism was looked upon as an independent and self-contained unit. This is understandable when we realize that man was surrounded by a bewildering variety of organisms which had to be studied one at a time. Naturalists had a formidable task in taking a census of life: in collecting, identifying, naming, and classifying all kinds of plants and animals, many brought from little-known distant lands. In so doing they eventually obtained an inventory of the earth's inhabitants, a catalog with more than a million entries, each describing a particular species. As this census neared completion, biologists were able to focus their attention more and more on the structure and functions of each organism, comparing the anatomy and physiology of one species with that of another. Thus they learned much about the self-maintenance, growth, response, and reproduction of each species.

But, while delving deeper and deeper into the inner secrets of nature, many biologists neglected to think of life in its total and collective aspect. It was assumed that a community was merely the sum of all the species found in a particular place. The vital relations of organisms to their physical environment and to each other were frequently overlooked or disregarded completely. Nevertheless, here and there a few observant naturalists were beginning to take note of such interrelations. Christian Sprengel, for example, in 1793 published an account of the intimate relationship that exists between insects and flowers. His revealing study was ignored by most of his contemporaries, perhaps because of the unscholarly but alluring title, *The Secret of Nature Discovered*. Charles Darwin was one who did take notice of Sprengel's work. He also made some keen observations of his own on life and environment, among them his well-known studies on the role of earthworms in conditioning the soil. By the end of the nineteenth century increasing numbers of biologists began studying the ways in which environment influences the structure, habits, and distribution of plants and animals.

This point of view became crystallized in 1870 when the German zoologist Ernst Haeckel coined the term "oekologie" for what he called "the total relations of the animal both to its organic and inorganic environment." He based the new term on two Greek words, *oikos*, meaning "home," and *logos*, meaning "study." Later the meaning of ecology (its modern English spelling) was extended to include plants as well as animals. Ecology comes from the same Greek root as economics,

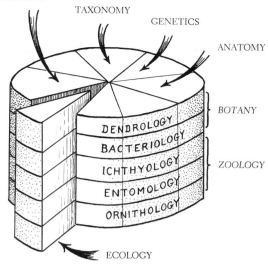

Biology can be compared to a layer cake, each horizontal section representing a bioscience dealing with a special group of organisms. The cake can be cut into slices that deal with fundamental aspects of life common to all layers. One such slice is *ecology*.

and in a sense can be called the "economics of nature." In defining this area of biology, Haeckel gave scientific status to what had been a vaguely amateurish pursuit popularly known as natural history. The study of organisms at home — i.e., in their natural surroundings — was given a great impetus by him.

Research in ecology has many far-reaching applications which are now beginning to be understood by industry, by government agencies, and by the general public. What effect will overgrazing in Nebraska have on a farmer who lives many hundreds of miles away? How will a forest fire affect the water supply of towns far distant from the devastated area? What influence will widespread use of insecticide sprays have on the pollination and fruit production of blueberries?

Ecology has many answers to practical problems in conservation, forestry, agriculture, and game management. Conservation is concerned with the control of dwindling water supplies, the prevention of floods and costly erosion of fertile soil, the avoidance of man-made droughts and forest fires. The wise use of all natural resources depends upon an understanding of the laws governing the interrelations of life and its environment. In agriculture, older trial-and-error methods are being supplanted by the more effective procedures based on ecological research. The use of chemicals as insecticides, herbicides, and pesticides is primarily an ecological problem. The widespread and careless use of chemicals injects new factors into the environment and alters the delicate balance which often exists in a wildlife community. Ecological knowledge is also helpful in avoiding modification of the environment by pollution of land and water. Ecology is the basis of forestry and the preservation of our wilderness areas and wildlife refuges, and in the preservation of natural areas under the jurisdiction of our federal, state, and local authorities.

This ecological story could well be entitled "Environment and Life: A Drama of Nature." At first man believed he could sit in the audience and view such a drama as an outsider, since it was his conviction that he stood apart from nature and could go on his own way with disregard of the laws that governed other living things. Now we know that we are far from passive members of the audience. In fact man is not only one of the actors in the drama but a very important member of the cast because of his unique ability to control the environment. Much of the past role played by Americans has been unfortunately one in which we can have little pride. Man's tampering with the environment has been marked by costly mistakes, and mistakes are continuing to be made

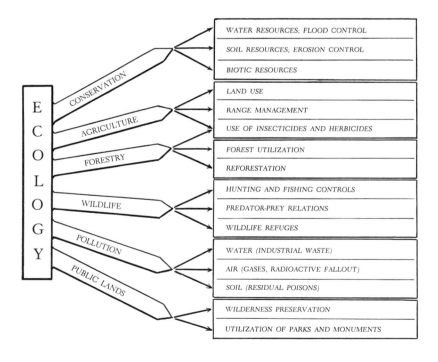

Ecology has far-reaching applications in many areas that affect our lives.

today. In the past these could be attributed to ignorance. In the future we have no such excuse.

Our American heritage is a fortunate combination of many different types of environment favorable to life, as well as many forms of wildlife that have evolved through millions of years of adaptation to these environments. The result provides a home for man which is perhaps unequaled elsewhere on earth. It is a priceless legacy which we can enjoy or abuse, conserve or squander. We stand today on the brink of possessing new sources of energy capable of contaminating, modifying, and even destroying much of this environment. Future generations will be faced with many difficult decisions, but foremost will be that of preserving this legacy in an age which combines atomic energy with a phenomenal increase in the human population.

CLARENCE J. HYLANDER

Bar Harbor, Maine
February 1964

Our way to the wilderness area takes us along an unpaved road into the heart of a wildlife community. (*Credit: George E. Nichols*)

I
Wildlife and Its Environment

1 / An Ecological Adventure

WHEN THE FIRST EUROPEANS landed on the shores of eastern North America, they found themselves on the edge of a vast forested wilderness some 400,000 square miles in extent. Dense growth of maple and beech, oak and hickory extended from the Atlantic Ocean to the Mississippi River. This wilderness provided food and shelter for an abundance of wildlife. Today, after a few centuries of human occupation, less than 2000 square miles of this forest remain in the original state. Some of these have become protected areas in state and national parks, forests, and preserves — fortunately saved from the destruction which has been widespread elsewhere. Scattered fragments survive in remote areas, spared from the saw and the bulldozer because useless for practical purposes. It is such a remnant of the original eastern forest which is the objective of our first ecological adventure.

Our way to the wilderness area takes us through a countryside where the original forest has given way to towns and cities, farms and cultivated fields. Leaving the well-traveled traffic lanes, we follow an unpaved road that winds through fields now abandoned and reverting to a tangle of weeds and grasses intermingled with sapling poplars and red cedars. These are the only trees the impoverished soil is able to support; seedlings from the neighboring virgin forest, which would be nourished were the soil in its original fertile state, can no longer dispute the invasion of the intruders. On entering the forest we at once notice a change in the surroundings. The air is cool and moist, the light less intense, the soil is rich and spongy. Tall trees soar upward to form a green canopy that screens out the direct sunlight. Patriarchs of the wilderness, these trees are century-old maples and oaks mixed with giant hemlocks and white pines. Beneath them grow smaller dwellers of the forest, dogwood and ironwood, which form a lower level of foliage in the shade of their taller associates. Nearer the ground is a still lower level of laurel and viburnum, and the forest floor is decked with dogberry, bunchberry, and Canada mayflower. A carpet of mosses and ferns spreads over fallen logs and boulders. Not a square inch of the earth's surface is devoid of vegetation.

Seated for a moment's pause on one of the moss-covered logs, we become aware of the many small animals that live amid the greenery close to the earth. Ants and beetles crawl through the leafy litter on the ground; grubs and worms burrow into the rotting logs. When we turn over a stone at our feet, a salamander wriggles from under it to seek a new moist hiding place. Beneath a nearby clump of ferns a wood frog sits motionless in the shade, while a sleek black snake slips noiselessly through the undergrowth in search of just such a meal.

More conspicuous are the birds that inhabit this forest world. A towhee scratches vigorously among the fallen leaves for its meal of insects or seeds. A downy woodpecker, inspecting the dead limb of a maple, drums noisily as it drills for larvae hidden beneath the bark. Chickadees flit from bough to bough, picking off insects we cannot see. Suddenly the woodland quiet is broken by the noisy takeoff from the bushes of a grouse that has been feeding on the seeds and fruits. A great horned owl, sitting high on a limb, is practically invisible because of its gray-brown feathers, which make it almost indistinguishable from the surroundings. But with the approach of dusk this winged hunter too will come to life and search the forest for the small mammals that make up its diet.

As we resume our way along the wood road we catch a glimpse of some of the larger mammals that make this wilderness their home. A porcupine shuffles into sight, nibbling at the tender foliage and chance fruits along its path. A raccoon ambles into view and pauses to search for acorns or ground-dwelling insects. Sensing our presence, the raccoon lifts its black-masked face to peer at us, then scrambles up the nearest tree. From the safety of its perch high among the swaying branches it eyes us inquisitively. From another tree a red squirrel chatters its annoyance at having been interrupted in the business of gathering seeds from the fallen pine cones.

Where the road passes through an open glade we catch sight of a cottontail rabbit partially hidden in a thicket. Sharing the feeding ground is a young fawn still spotted with white. Motionless and alert, its keen ears have picked up the sound of our movement and, after a breathless moment, the deer bounds away with a flash of white rump. Farther on, at a bend in the road, we see another elusive forest dweller, a red fox trotting briskly yet quick to spy an unwary mouse or cottontail. The wilderness trip would be complete if a bobcat, the uncontested ruler of this domain, were to appear. But during the daytime this wary predator is undoubtedly stretched out on a stout limb, waiting like the owl for the good hunting that comes with the dusk.

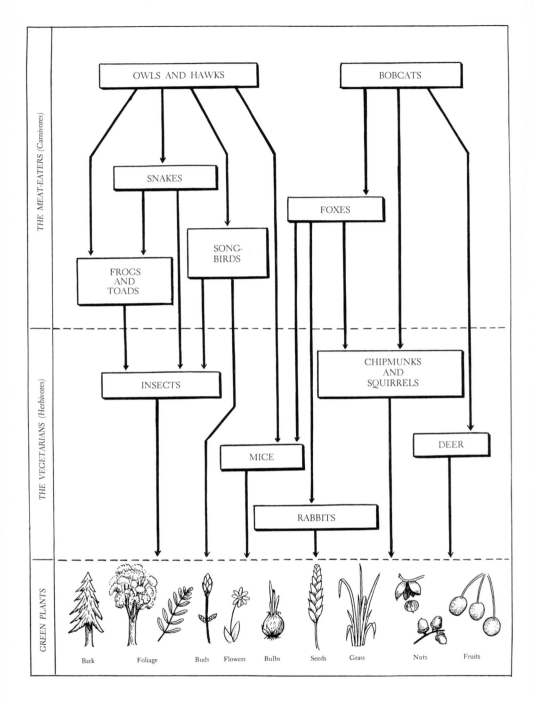

Even a small wildlife community is a complex web of food relationships among its members.

In other areas, also, the forest reveals its inhabitants as we pass. A swampy pond, rimmed with cattails and sedges, is the haunt of red-winged blackbirds and blue herons. The telltale pile of sticks and grasses rising above the surface of the water is the home of a muskrat family. Prowling among the alders that surround the pond, a hungry weasel vigilantly waits for a careless muskrat, its favorite meal. Farther along, the road leaves the swampy ground and winds up through a rocky ravine whose slopes are covered with smooth-barked beeches and lacy hemlocks. A sleek chipmunk, cheek pouches stuffed with acorns, darts from stone to stone on its way from an oak tree to its cache deep in the rocky labyrinth.

The road now reaches the far side of the wilderness area, the shady tunnel of green widens as it leads into the blinding sunlight and oppressive heat. The weedy fields seem devoid of life in contrast to the forest through which we have just passed and within whose shelter and protection we have seen such varied wildlife.

There are many ways of looking at such a vignette of nature. To a casual hiker it may be a restful escape from the noises and odors of the city. He undoubtedly considers the woods a pleasant assortment of trees and insects and birds. It is entirely possible that, being unfamiliar with the identity of particular plants and animals, he thinks of the woods as an assemblage of big trees and little trees, brown birds and gray birds, crawling bugs and annoying mosquitoes. He might even be dismayed by the face-to-face meeting with a porcupine or a black snake.

A second, more observant, approach characterizes the naturalist. His interest, however, is often limited by his particular hobby. If he is a botanist he is able to identify the different kinds of birches and maples, and his trained eye is quick to locate a rare orchid or violet; yet he may scarcely notice the frogs and the chipmunks. If he is a birdwatcher he may be so intent upon identifying the hairy woodpecker that he does not notice it is perched on a dead maple rather than a living birch. If he is an entomologist he finds the woods a paradise of beetles to be added to his collection, but the trees and the birds may arouse little interest. Being intent upon isolated segments of nature, such naturalists often miss the complete picture of the woods as a community of plants and animals. This latter is the broad outlook of the ecologist.

From the ecological point of view, the forest cannot be described as an arithmetical equation of 25 trees + 20 birds + 10 mammals = 55 kinds of wildlife = the forest. Instead, the forest is considered a living unit within which definite functions are assigned to certain types of plants and animals. Such a unit of nature, consisting of organisms living together in the same environment, is known as a *biotic community*. This community has a definite membership, it is characterized by definite physical living conditions, and consists of an internal organization with mutually dependent organisms forming a complex "web of life."

The concept of the forest as a community may answer many questions that have come to our minds as we explored the wilderness area. Why is this area woodland instead of grassland? Why are the woods made up of maple and oak, instead of spruce or palmetto? Why are there mammals, raccoons and chipmunks instead of opossum and kangaroo rats? These and many additional "whys" can be answered by considering the physical environment and its influence on the membership of the community.

A major factor is climate, which determines weather conditions, temperature, and rainfall. The precipitation in the region we have explored is about 45 inches a year, some as rain and some as snow. This amount is adequate for the growth of trees; thus the area is a forest. If it had been in eastern Nebraska, where the rainfall is less than 15 inches a year, the vegetation would have been grasses and not trees, since trees do not grow readily where the annual precipitation is less than 20 inches a year. The temperature in our forested area ranges from zero to 100° F. with seasonal variation from summer to winter. This periodic temperature change requires a special adaptation on the part of some trees which cannot carry on their food-making activities during the winter. As a result these trees are deciduous, losing all their foliage at once and relapsing into a dormant leafless state in winter. The forest in the region we explored includes trees adapted in this way, such as maples and oaks. If the woods were a thousand miles farther north, the more rigorous living conditions would have eliminated these deciduous trees and the forest would be an evergreen one of spruce and fir. Such evergreen trees are adapted to retain their foliage throughout the year and are characteristic of cold climates. If, on the other hand, we had chosen a woodland a thousand miles farther south, the milder climate would encourage the growth of other deciduous trees, such as tuliptree and sweetgum, and broad-leaved evergreens like holly and magnolia.

Climate likewise determines the types of animal life which occur in a community, and their living habits. If our forest wilderness had been located in a southern region, the community would include a larger population of amphibians and reptiles, since these animals are not adapted to live in cold climates. The birds would be cardinals and mockingbirds instead of chickadees and ruffed grouse. The mammals would include opossum and fox squirrels instead of porcupines and red squirrels.

Other physical factors vary over different parts of the same forest: the amount of light, the kind of soil, the amount of water in the soil. These factors determine the distribution of plants within the community. Some plants demand considerable sunlight, others can grow in weaker light, still others can grow even in the dense shade of the forest floor. The canopy trees — the maples and birches and oaks — are sun-lovers and compete with each other for a place in the sun. In their shade grow trees that can tolerate dim light such as dogwoods and ironwoods. Mosses and ferns live close to the ground, since their delicate leaves require shade and moisture, conditions provided by the dense layers of taller vegetation. Soil conditions vary from deep spongy humus-rich soil in the hollows and along stream margins to shallow rocky soils of ravine slopes and ridges. Birches and maples thrive in the former soils, pines and oaks in the latter. The amount of water retained by the soil also varies within the forest, and is responsible for the distribution of the vegetation. In wet locations grow plants adapted for living with their roots in water-soaked ground. In such locations we found cattails, red maples, and alders. In the mediumly moist soil of the woods, birches and beeches find satisfactory homes. Hickories and oaks grow on the higher, drier sites like ridges and hilltops. These factors influence animal life less than plants, since animals are not rooted in one spot but can move about in search of water and suitable light conditions.

The concept of the forest as a community also provides answers to other questions about the living environment. What would happen to the trees if woodpeckers were not present to eat the bark insects? If all the oaks should die, what would be the effect on the chipmunks? If the bobcats should be exterminated would this have any effect on the seedling birches and maples? Why are amphibians most abundant in swampy locations, and what would happen to them if the swamps dried up? What determines the relative numbers of rabbits and deer in the woods? These are some of the questions we shall attempt to answer. But first we must meet the various types of inhabitants that live in a wildlife community.

2 / An Inventory of American Wildlife

THE LIVING ENVIRONMENT is a combination of three types of organisms: green plants, animals, and nongreen plants. The green plants, being the food producers of the community, are an essential part of the living environment. Since they remain in one spot throughout their lives, they form the vegetation which gives a permanent aspect to every wildlife area. Animals are dependent upon green plants for food and shelter; thus their distribution is determined by the vegetation. Being able to move about, animals are often a less permanent feature of a community, and a less conspicuous one. The nongreen plants are vitally important to community life, even though they are usually unobtrusive. They include the bacteria, molds, and mushrooms. Being unable to carry on photosynthesis, they too are dependent upon green plants for the ultimate source of their food.

THE GREEN PLANTS

In our exploration of American wildlife communities, we shall find that the important green plants are the trees and shrubs, grasses and other herbaceous plants, mosses and lichens. These are classified into two groups on the basis of whether they reproduce by means of seeds or spores. The *seed plants* are the larger plants of the community and typically develop roots, stems, and foliage. The seed is a highly specialized reproductive organ containing an embryo plant. It additionally contains a supply of stored food for the use of the developing seedling; this food is also a staple diet of many animals. Some seed plants have woody stems; these are the trees and shrubs. Others lack wood and are known as herbaceous plants; they include the grasses, cacti, and the many small flowering plants. The *spore plants* are the smaller plants of the community, and most of them lack roots and specialized leaves. The reproductive structure is a single-celled spore which, containing no stored food as does a seed, is of little food value to animals. Mosses and lichens are the simplest of the spore plants; ferns are more highly developed, with true roots and leaves.

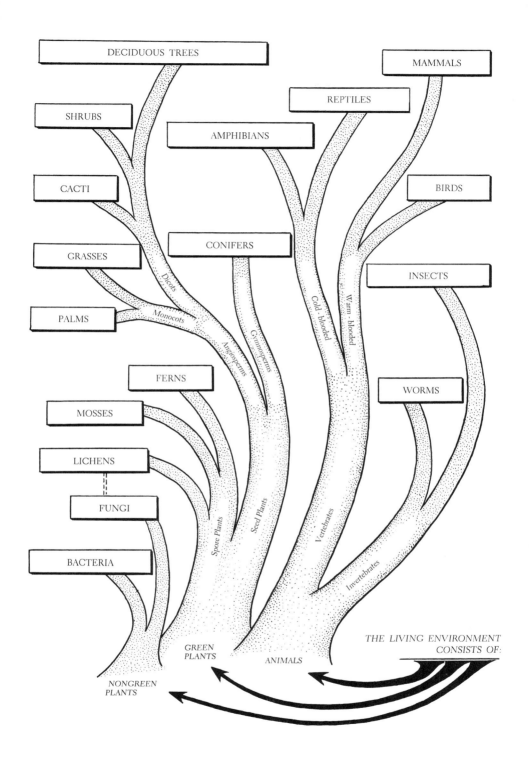

The wildlife community is a combination of three types of organisms: green plants, nongreen plants, and animals.

16 / WILDLIFE AND ITS ENVIRONMENT

Seed plants make up the major part of the vegetation of forests, grasslands, and deserts. They have differences in foliage and method of seed production which can be used in classifying all seed plants into two groups: gymnosperms and angiosperms. Each group has features that fit it for living in certain types of environment, and for supporting special kinds of animal life. *Gymnosperms* (from the Greek words meaning "naked seeds") produce their seeds in cones, and each seed is shed without any protective covering. The leaves are usually small and evergreen. Common gymnosperms are the conifers, of which pine and spruce are examples. Conifers thrive on high mountains and in the

Gymnosperms produce their seeds in cones, and each seed is shed without any protective covering. Angiosperms produce seeds by means of flowers, and the seed is enveloped in a fruit.

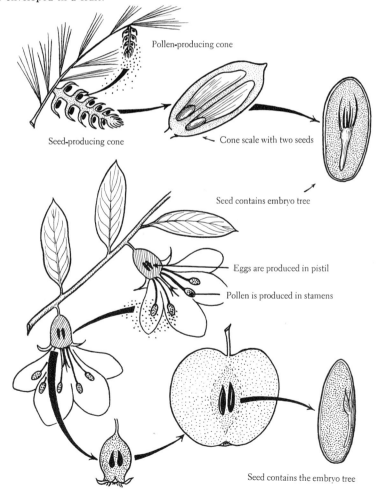

Pollen-producing cone

Seed-producing cone   ← Cone scale with two seeds

Seed contains embryo tree

Eggs are produced in pistil

Pollen is produced in stamens

Seed contains the embryo tree

Fruit forms after pollination

northern portion of the continent. Conifer foliage is for the most part too harsh and inedible for forage, but the seeds are prized by many birds and mammals. *Angiosperms* (from the Greek words meaning "seeds in a container") produce their seeds by means of flowers, instead of cones, and the seed is enveloped in a protective fruit. In the majority of species, the broad, thin leaves are shed at the onset of a cold or dry season. Thus angiosperms make up our deciduous forests. Foliage of angiosperms, as well as their seeds and fruits, provide most of the diet of herbivorous animals. Angiosperm trees thrive in less rigorous environments than conifers, and so occur lower on the mountain slopes and in the temperate portion of our continent.

Conifers have two types of foliage, which offers a convenient means of separating them into two main groups. One has long, slender, needlelike leaves; this is typical of pine, spruce, fir, and coast redwood. The other group has small, scale-like, and overlapping leaves; this type is found in juniper, cedar, and Sierra redwood. Of the needle-leaved conifers, two — the pine and the larch — produce their needles in clusters. Pines include a great number of species in North America. One group of pines bears five needles in a cluster, as is true of the white pine. Another group has three needles in a cluster, as is the case with ponderosa pine. A third group, of which lodgepole pine is typical, has two needles in a cluster. Larches, also known as tamaracks, have needles in clusters but these form star-shaped rosettes of a dozen or more short needles. Larch needles are not evergreen, hence this tree is one of the few deciduous conifers.

The remaining needle-leaved conifers bear leaves singly, either on all sides of the twig or in two rows that form a flattened foliage spray. In the spruces the squarish needles surround the twig, and produce stiff and prickly foliage. Firs bear flattened needles in two opposite rows along the branches. Hemlocks also have flattened needles in two ranks along the branches, but the needles are smaller in size and form a more lacy foliage. Two of the flattened-needle type of conifer are unique members of their respective communities. One is the coast redwood, found only along the coast of Oregon and California. The other is the bald cypress, which forms dense swamp forests along the southeastern coast. Bald cypress, like the larch, is deciduous.

Of the conifers with scale-like leaves, the junipers are usually shrubs or small trees; a common eastern species, called red cedar, thrives in open sandy fields. White cedar and arbor vitae are partial to cold swamps and low wet ground; their foliage provides winter forage for

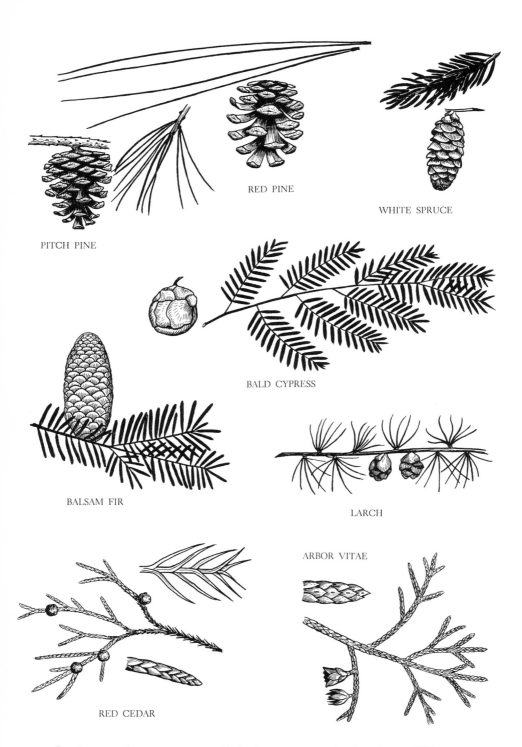

Conifers are the most common kind of gymnosperms in American wildlife communities.

many of the forest-dwelling animals. The true cypress, with foliage and cones similar to those of the white cedar, is found only in the western mountains and on the Pacific Coast. The best known of the scale-leaved evergreen group is the Sierra redwood, also known as the "big tree." This patriarch of all conifers grows only in the California mountains.

Angiosperms are classified into two groups on the basis of their seed structure. In one group the seed develops two primary seed leaves, or cotyledons; the group is therefore known as dicots (meaning two cotyledons). In the other group the seed develops one cotyledon, and the group is known as the monocots (meaning one cotyledon). Dicots usually have a distinctive netted-veined leaf pattern, while the leaf of monocots is parallel-veined. Common dicots are maples and oaks, common monocots are palms and orchids. The two large subdivisions of the angiosperms, therefore, can be recognized by their leaves as well as by their seeds. The dicot trees that we shall encounter in our ecological exploration of North America fall into three categories: the catkin-trees, the trees with inconspicuous flowers, and the trees with conspicuous flowers. The monocot trees form a fourth group; they are the palms and yuccas.

Catkin-Trees. A catkin is a cone-like reproductive structure made up of many small greenish or yellowish flowers arranged along a central axis. The "pussy" of a pussy willow is a typical catkin. The catkin-trees include the oaks, the nut trees, the poplars and willows, the birches and alders. Most of these trees are deciduous; they make up much of the vegetation of eastern forests, and provide food and shelter for many birds and mammals living in these communities. Oaks, of which there are many species, are found in every part of the United States. The pollen-producing catkins form pendent tassels, the clustered seed-producing catkins develop into acorns. Some oaks are deciduous, like the eastern black oak and white oak. Others have evergreen foliage; these are the live oaks of the southern and western states. The nut-bearing trees, the walnuts and hickories, are an important source of food for many woods mammals. Other nut-bearing trees are the beech and the chestnut, once abundant throughout the Appalachian Highlands. Poplars and willows have both the pollen- and the seed-producing catkins in elongated tassels. Poplars are fast-growing trees that often form thickets in open clearings; some species are known as aspen and cottonwood. Few deciduous trees have become adapted to

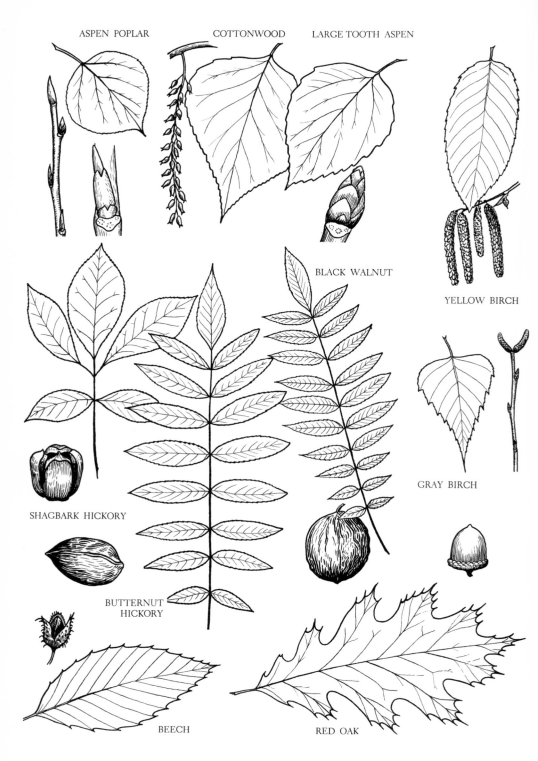

Oaks, nut trees, birches, and poplars are familiar angiosperm trees reproducing by catkins.

the arctic regions. However, willows have the distinction of being able to live farther north than any other woody angiosperm, and their foliage provides forage for many arctic herbivores. Birches and alders include many hardy species; the white birch is a companion of conifers in northeastern evergreen forests, and alders grow in the Far North. Alders, with a preference for stream margins and swamps, provide cover for many of the birds.

Trees with Inconspicuous Flowers. The second group of angiosperm trees reproduces by flowers which, although not in catkins, are equally small and inconspicuous, often greenish or yellowish. Some of these trees are important members of the forest community. Typical are the maples and sycamores. Maples are found throughout the eastern deciduous forest. Sugar maple has yellowish blossoms; red maple, also known as swamp maple, has reddish blossoms. Stream margins and moist canyons, in addition to woods, are the homes of sycamores, also known as buttonball trees in reference to their pendent spherical fruit.

Trees with Conspicuous Flowers. The third group of angiosperms produces colorful and conspicuous flowers. Their fruits are often large, fleshy, and edible and thus important in the diet of many birds and mammals. Most representative of this group are the rose and legume families with many members which play a significant role in the food relations of the community. The rose family includes the spiny-branched hawthorns, small trees with juicy red fruits, as well as many different kinds of wild cherries, plums, and apples. Various trees of the legume, or pea, family grow in the arid southwestern states, where they provide food for desert animals. Other trees with conspicuous flowers are chiefly eastern in distribution: the buckeye, a relative of the cultivated horse chestnut with similar showy flower cluster; the magnolia, of which some species are evergreen; the dogwood, an undergrowth tree with showy white blossoms and red berries for fruit; and the tall tulip-tree, with large yellow blossoms.

Monocot Trees. The last group of trees are those with parallel-veined leaves. They include the palms and yuccas. Palms are tropical plants poorly adapted for life in cold habitats. The tall unbranched trunk of a palm bears a crown of huge leaves; the summit of the trunk bears clusters of the fruit, which may be berries or larger fruits such as dates and coconuts. The cabbage palm, an eastern species, forms dense jungles in the swampy lowlands of the Atlantic Coastal Plain. The Washington palm, or desert palm, grows along watercourses in the southwestern desert. Related to the palms are the succulent-leaved

Palms are monocot trees with tall unbranched trunks and a crown of huge leaves.

yuccas, whose stiff sharply pointed leaves have given the shrubby species the common names of Spanish dagger and Spanish bayonet.

In addition to the angiosperm shrubs and trees, there are two groups of herbaceous angiosperms which are important in an ecological exploration of our continent: the grasses and the cacti. *Grasses* belong to the monocot group of angiosperms, as their long, narrow, and parallel-veined leaves indicate. The many different species of grasses, from bluegrass to buffalo grass, have rather uniform foliage characteristics; it is in the flower structure that their chief differences are apparent. Grass flowers are small, greenish, or yellowish wind-pollinated flowers, growing in terminal spikes or clusters. The seeds, though small, are an important source of food for numerous birds and small mammals. Grass foliage provides forage for all our herbivores from mice to mountain goats. Grasses are able to live in regions where the rainfall is insufficient for tree growth; thus they are the dominant vegetation in the semi-arid great central lowlands of the continent. Grasses spread rapidly by underground stems and perennial roots, and thereby can monopolize the ground, preventing invasion by seedlings of shrubs and trees.

AN INVENTORY OF AMERICAN WILDLIFE / 23

Cacti belong to the dicot group of angiosperms. The majority are small plants, although a few, like the shrubby prickly pear and the saguaro, attain the stature of small trees. They are the dominant vegetation of many parts of the arid Southwest, where they provide shelter and nesting sites for many desert mammals, reptiles, and birds. Contrary to public belief that cacti are exclusively desert dwellers, they also grow in a variety of other habitats. Species occur throughout the prairie states to Canada, in the mountains, and amid the lush grasses of the Atlantic Coastal Plain. Some cacti have jointed stems made up of flat sections, as in the prickly pears. Others have cylindrical jointed stems, as seen in the chollas. Still others have compact hemispherical

Cacti are leafless angiosperms with enlarged green stems that carry on photosynthesis.

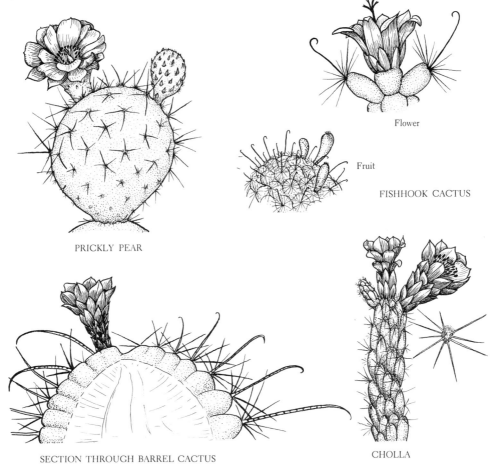

PRICKLY PEAR

Flower

Fruit

FISHHOOK CACTUS

SECTION THROUGH BARREL CACTUS

CHOLLA

or cylindrical bodies, common among the pincushion and barrel cacti. Largest American cactus is the saguaro of the Southwest.

The *green spore plants* are for the most part small and insignificant, but two kinds are of importance in the wildlife community: the mosses and the lichens. Mosses lack roots and have very primitive, small leaves. Most mosses require shaded and moist habitats, where they form a compact ground cover, spreading over rocks and fallen trees. A few mosses live in water and wet depressions; such a moss is sphagnum, also known as peat moss. Other mosses, like the brownish-green rock mosses, are able to grow in rock crevices exposed to bright sunlight. When other food is not available, mosses are eaten by arctic herbivores. Lichens are grayish-green plants with even simpler body structure than mosses. Although they superficially resemble mosses, and like them have no roots, lichens are not differentiated into stems and leaves. Theirs is a simple type of plant body known as a thallus. Lichens can grow on bare rock and earth, and so can live in many habitats unsuited to other plants. Some lichens are thin and papery, forming crusts on rocks and cliffs. Others grow into erect cushions, such as the reindeer "moss," which is so important in the diet of caribou; and others are air plants, decorating the branches of trees in cool damp forests of the North. Lichens are the hardiest of pioneers, growing farther north and higher above timberline than any other kind of plant. They can do this because of their total independence of soil.

The Cold-blooded Animals

The animals found in our American wildlife communities are so numerous and varied that at first sight they may seem a bewildering assemblage. But, as with plants, we can identify them by a classification system developed by biologists. Such a system separates the animal kingdom into two major groups: invertebrates and vertebrates. The *invertebrates* are those animals which either have no skeleton or have the skeleton on the outside of the body. Many invertebrates live in the sea and are small and relatively simple animals. Others are an inconspicuous part of life on land: the worms, snails, millipedes, and spiders. One group — the insects — are very important in every wildlife community. Some invertebrates are parasites that play a major role in the interrelations of wildlife. *Vertebrates*, animals with an internal skeleton of bone, are more conspicuous and familiar members of the wildlife community. Of these, four groups live on land: the amphibians, reptiles, birds, and mammals. All in common have a backbone or vertebral column as a central axis to the body. This type of construction permits a vertebrate to grow to larger size than can most invertebrates, with the possibility of more complex behavior. Each of the four groups possesses characteristics that adapt and sometimes restrict them to special environments. Of these, the type of body covering and the method of reproduction are used as a basis for classifying vertebrates into smaller groups.

Vertebrates can be divided also into two groups on the basis of their ability to maintain a constant body temperature. Some land vertebrates, such as the amphibians and reptiles, are *cold-blooded*, that is, their body temperature varies with that of the environment. On a cold or cloudy day their body temperature is lowered, on a sunny warm day their temperature rises. Other vertebrates — the birds and mammals — are *warm-blooded*, being able to maintain the same body temperature regardless of how much the temperature of the surroundings may vary. We shall see later how this ability to keep the body at a constant temperature is of advantage to animals living in cold climates.

Most primitive of the land vertebrates are the *amphibians*, whose thin unprotected skin cannot prevent evaporation of water from their bodies. Therefore they are restricted to shady, damp habitats. Amphibians lay their eggs in water, where the young (tadpoles) develop. For this reason, amphibians have to live near ponds or streams. Newts and salamanders are amphibians with elongated bodies, long tails, and short stubby limbs. Other amphibians — the frogs and toads — have more compact bodies, with longer, more muscular limbs; they lack tails when adult. Amphibians are chiefly carnivores, feeding on insects and insect larvae, worms, and other small invertebrates.

Reptiles are better suited for life on land since they have a thicker skin protected by scales. Because they lay their eggs on land, reptiles are independent of an aquatic environment. As both amphibians and reptiles are cold-blooded, these groups are not common in the colder portion of our continent. American wildlife communities include four groups of reptiles: lizards, snakes, turtles, and crocodilians.

Lizards have elongated bodies with a flexible armor of small overlapping scales. Their long tails are fragile, often breaking off when grasped by a predator. Most lizards are carnivores, feeding on insects, worms, and smaller reptiles. Our lizard species include skinks, fence lizards, anoles, horned lizards, chuckwallas, and Gila monsters. Skinks can be distinguished by their glassy-smooth small scales. A number of different kinds of fence lizards, with rough-scaled bodies, live in open sunny fields; they are adept at scurrying over fences and climbing tree trunks. Best-known lizard of the southeastern United States is the anole, also called the American chameleon because of its ability to change color from brown to green. The horned lizard, incorrectly called a "horned toad," is a grotesque little short-tailed reptile of exceedingly hot habitats. Largest of the American lizards are the chuckwalla and Gila monster; the chuckwalla, an iguana-like lizard of rocky deserts, is one of the few herbivorous reptiles and the Gila monster is our only poisonous lizard.

Snakes unfortunately evoke an unreasoning fear in many persons, preventing them from thoroughly enjoying the exploration of the out-of-doors. Snakes are actually secretive and nonaggressive members of the wildlife community, with interesting habits. They are much like lizards in body structure, and possess the same type of body armor made of small flexible scales. The forked tongue, popularly mistaken for a stinging organ, is important in detecting chemical and physical stimuli. A distinctive snake feature is the hinged lower jaw making it possible for a snake to swallow food in huge pieces, sometimes to devour a small mammal whole. All snakes are carnivorous; their diet consists of insects and other small invertebrates, amphibians, other snakes, birds and bird eggs, and rodents. A preference for rats and mice makes snakes valuable allies in curbing the increase of these destructive herbivores. A number of harmless snakes live in the forests and grasslands, but they are rarely seen because of their unobtrusive habits. Some snakes have special hollow teeth known as fangs through which venom flows from sacs behind the eyes. Smallest of these venomous snakes is the brilliantly colored coral snake, ranging from the southern United States to Argentina. Other fang-bearing snakes belong to the pit-viper group, so named because of a heat-sensitive organ between the eyes and nostrils which enables the snake to detect the presence of warm-blooded prey. Rattlesnakes, cottonmouths, and copperhead are American pit vipers.

Turtles number fewer species than either lizards or snakes. In popular language, the name turtle is restricted to aquatic species, tortoise to the land dwellers. A turtle's distinctive shell is made of fused plates, attached permanently to the backbone and ribs; the shell takes the place of the flexible body armor of other reptiles. With such protection, turtles rely upon passive resistance when attacked. Most turtles are omnivorous; the common spotted turtle is among the few strict carnivores. Even fewer species are herbivores, as, for example, the gopher tortoise. Turtles are particularly abundant in swampy deciduous woodlands of eastern United States; here one can meet painted and spotted turtles in ponds, wood turtles and box turtles in more terrestrial habitats. The gopher tortoise is the most confirmed land dweller of the group, burrowing into the sandy soil of Florida and hiding among the rocks of the southwestern deserts.

Crocodilians are amphibious reptiles that add considerable interest to the wildlife communities of the southeastern United States. They are our most aggressive reptiles. Commonest is the blunt-nosed alligator, a powerful reptile of the cypress swamps of Georgia and Florida. Alligators feed on any kind of prey they can capture, from fishes, amphibians, other reptiles, and mammals to birds such as herons and egrets. The American crocodile has a much more restricted range, living in the brackish and saltwater mangrove swamps of the southern tip of Florida.

The Warm-blooded Animals

Birds are such familiar members of our wildlife that they hardly need an introduction. In a number of species, they exceed all other groups of land vertebrates. Most birds are small, with a weight measured in ounces rather than pounds. But some species have an impressive wingspan; that of a condor is almost ten feet. The majority of birds are omnivorous, eating either insect or plant food, depending upon its availability. A few are strict carnivores, such as woodpeckers and birds of prey. Some carnivores — vultures, crows, and gulls — have become scavengers. Game birds are among the few strictly herbivorous species. Birds of ecological importance fall into four groups: perching birds, game birds, water birds, and birds of prey.

About one third of the common birds are perching birds. Many of these are songbirds, adding sound as well as color to the wildlife communities. In the perching birds, three of the four toes are directed forward, one backward — a valuable adaptation for clinging to branches. Birds show definite preference in choice of both temperature and habitat. Snow buntings and Lapland longspurs live under arctic conditions. Chickadees and jays prefer cool conifer forests. Vireos and warblers are numerous in eastern deciduous forests, meadowlarks and goldfinches in fields. Even in the seemingly inhospitable desert, cactus wrens and orioles have their homes. Perching birds are important in the economy of nature because of their appetite for insects, a natural check on these insatiable herbivores.

Game birds are relatively large herbivorous birds whose strong toes are adapted for scratching for seeds and fruits among fallen leaves. Al-

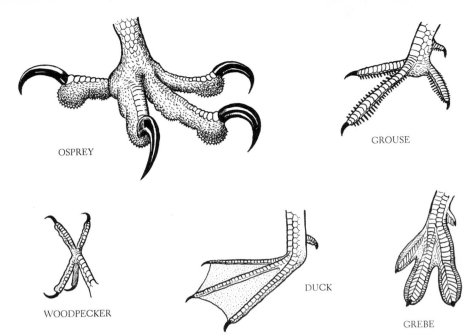

The living habits of birds are revealed by the special adaptations of their feet.

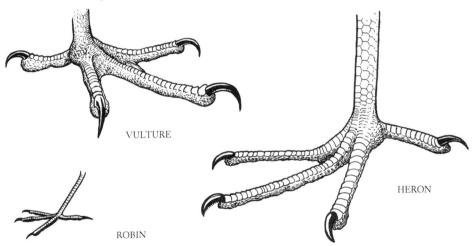

though able to fly, most of them resort to running or rely upon camouflage to escape enemies. Because they provide both food and sport, a great number of game species are threatened with extinction. Like the perching birds, game birds have distinct preferences in climate and vegetation. The rock ptarmigan is a hardy bird of arctic regions; the ruffed grouse frequents cool conifer woods, and the bobwhite quail lives in grassy woodlands. The largest game bird is the turkey, formerly abundant from New England to Florida.

Water birds rely upon aquatic plants and animals for food. They are

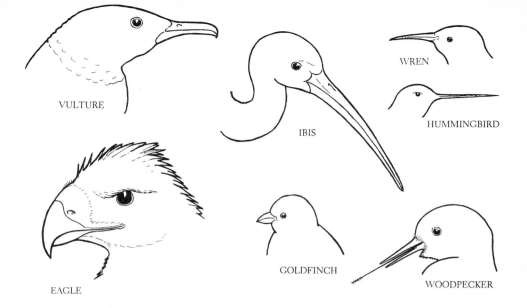

The living habits of birds are also revealed by the modifications of their bills.

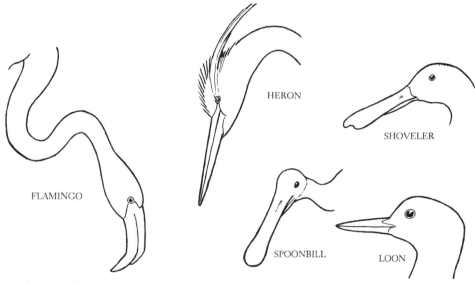

abundant in the Arctic during the summer breeding season, and are found throughout the year in all other parts of the continent. The loon is a diving bird, typical of northern lakes, and the anhinga is a bird of cypress swamps of the South. Both are expert swimmers and feed on fishes or other aquatic vertebrates. Wading species, like the herons, are equipped by long legs to patrol shallow waters and impale frogs and fishes on their pointed bills. More southern in range are the white egrets and ibis, spectacular birds of Florida swamps. Ducks and geese migrate south to warmer climate in winter; while traveling from one home to another they follow definite flyways. Consequently, many communities at certain times of the year include numerous migratory species.

Birds of prey are the feathered predators, feeding on other birds as well as on all other groups of vertebrates. Their strong hooked beaks, sharp talons, and unusual power of flight place them in a top position in a wildlife community. Over two dozen species of hawks occur from coast to coast; they keep in check the prolific rodents and other small herbivores which threaten the food producers of the community. Owls, of which there are many species with a coast-to-coast range, are the hunters of the night. They too feed mainly on rodents. Largest of the carnivorous birds are the eagles and osprey hawks. Many of these birds of prey are near extinction because of man's misconception of their role in nature. Like all predators, they are often needlessly destroyed.

Mammals, like birds, are so familiar to the woodsman and naturalist that they need little introduction. Although comparatively few may be found in any wildlife community, they add immeasureably to the pleasure of exploring the outdoors. Mammals vary considerably in size, from shrews weighing a fraction of an ounce to bull bison weighing a ton. The land mammals of North America differ from one another in the ways their limbs have become adapted for locomotion and their teeth for obtaining food. Four kinds of mammals are common in the wildlife community: the primitive mammals, the gnawing mammals, the hoofed mammals, and the clawed mammals.

North America possesses few primitive mammals, but those that do occur are of considerable biological interest. These are the marsupials and the insectivores. *Marsupials* reproduce in a primitive fashion: the young are prematurely born and continue their development in a pouch of the mother's body. Marsupials are most numerous in Australia, where the best-known representative is the kangaroo. Our only marsupial is the opossum, an animal of eastern deciduous forests which feeds on insects, worms, berries, and fruit. The *insectivores* are small mouse-like mammals with primitive teeth; this group includes the shrews and the moles, both widespread in all areas but rarely seen because of their subterranean habits. Insectivores feed mainly on insects and other small invertebrates.

The gnawing animals are highly specialized for an herbivorous diet. They comprise an exceedingly large group of mammals; *rodents* and *rabbits* are familiar examples. Their incisor teeth are large, functioning

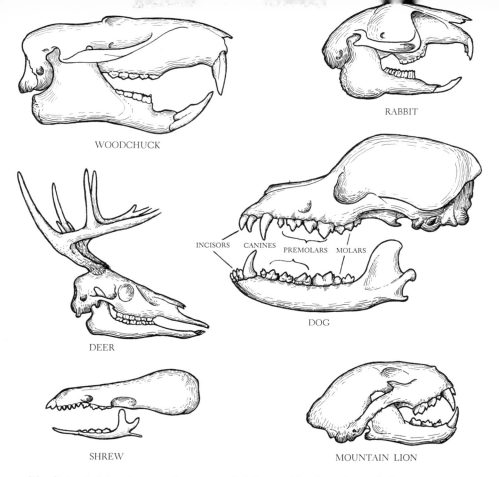

The living habits of mammals are revealed by special adaptations of their teeth.

as self-sharpening chisels which can gnaw through the trunks of trees and the toughest of seed coats. Grassy open areas are the preferred home of many rodents who are expert burrowers like woodchucks and prairie dogs. Mice, voles, and rats thrive in all types of habitats: jumping mice live in the cold northern forests, rice rats in the swamps, kangaroo and pack rats are adapted for life in hot arid regions. Forests are the home of many different kinds of expert climbers, such as squirrels and chipmunks; here too lives the spiny-armored porcupine, and the beaver. In the Far North the most abundant mammal is the prolific lemming. Closely related to the rodents are the rabbits and hares, distinguished by their overdeveloped hind limbs and jumping type of locomotion. These gnawing mammals range widely over the country, thriving wherever there is grass or other succulent plant food.

In the hoofed mammals, or *ungulates*, the toenails are modified to form a hoof, an adaptation for running rapidly over open ground. The teeth of hoofed mammals are modified for their particular diet of grasses and foliage; the molar teeth are specialized for grinding vegetation into digestible particles. Another characteristic of the group is the development of horns. The horns are of solid bone and shed each year in the deer family, which includes elk, caribou, and moose. The horns are permanent in the cattle family, with a hollow sheath fitting over a bony core; in this group are bison, muskox, mountain goat, and bighorn sheep. The pronghorn antelope likewise has horns consisting of a hollow sheath over a bony core, but the sheath is shed annually. Many of the hoofed animals provide man with leather, fur, clothing, and food; hence they have been almost completely exterminated in many parts of the country.

The clawed animals have specialized canine teeth instead of broad grinding molars. Their ability as predators is increased by a lithe muscular body, powerful limbs, and padded feet. In the *weasel family*

The living habits of mammals are also revealed by the modifications of their limbs.

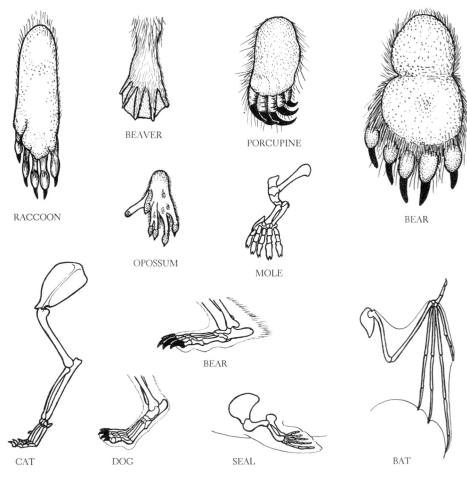

are several small, supple-bodied predators. The conifer forests are the home of such large species as wolverine and marten. Other members of the family include mink, prized for its fur coat, and the skunks, also well known but not valued so highly. The *cat family*, characterized by retractile claws, is a group of strict carnivores; in many communities it is the ruling dynasty among the predators. Our native "cats" include bobcat, lynx, mountain lion, jaguar, and ocelot. Widest in distribution is the bobcat, or wildcat, a medium-sized carnivore of forested and mountainous regions. Largest member of the cat family is the mountain lion, which goes by a number of aliases: cougar, panther, puma. Members of the *dog family* are often more omnivorous than carnivorous. In this family are the fox, coyote, and wolf. The common red fox ranges from northern Canada to the prairies and southern woodlands. It is best known of our large mammals because of its ability to survive in competition with man. The coyote, another mammal successful in living in proximity to man, has increased its range from the central prairie states both eastward and westward. Largest and rarest of the family is the gray wolf, today finding refuge in the remoter parts of Alaska and Canada. Most impressive of all our large mammals are the *bears*. They also are more omnivorous than carnivorous, feeding on honey, berries, and foliage as well as on insects and small mammals. The common black bear is found in dense woods from Alaska and Canada throughout the Appalachian highlands to Florida, and in the western mountains. Of more restricted range is the vanishing grizzly bear, an animal of the high western mountains and the northern tundra.

The green plants and the animals make up the most conspicuous portion of every wildlife community. The nongreen plants are inconspicuous and easily overlooked. They do, however, play an important role in the interrelations of organisms. The nongreen plants reproduce by spores instead of seeds and have simple bodies that do not develop roots, stems, or leaves. Since they lack chlorophyll they must rely upon other organisms for food, but unlike animals the nongreen plants obtain this food by absorption rather than by eating. Some feed on organic material while it is still a part of the living organism. These are known as *parasites*. Many of the plant parasites are the bacteria that cause animal diseases, others are fungi, which attack green plants. Another group of nongreen plants feeds upon the organic material left in dead plants and animals. These are known as *saprophytes*. Some of them, like the bracket and shelf fungi, are common growths on the trunks of trees. Others are the molds and mildews that grow on decaying vegetation and animal debris. Still others are the mushrooms that appear in leafmold and litter on the forest floor during damp or rainy seasons. Saprophytes and parasites are relatively uncommon in the cold Arctic, and increase in numbers in the more humid and warmer life zones of the southern states.

This is a brief inventory of the types of plants and animals we shall encounter as we explore the wildlife communities of North America. It emphasizes the variety of types of organisms rather than that of individuals. We shall now see how these different types are adapted to the physical environment which is their home.

3 / Adaptation of Life to Its Physical Environment

WE UNCONSCIOUSLY ASSOCIATE a plant or animal with a particular kind of environment. Spruces remind us of snow-covered northern mountains whereas cacti are the symbol of the desert. Prairie dogs call to mind vast reaches of prairie, and alligators have a moss-draped cypress swamp as a backdrop. It is common knowledge that the physical environment is somehow responsible for the characteristics and distribution of life. But why can a cactus grow where a spruce cannot, why can a prairie dog live where an alligator would die? For answers we must examine the environment more closely to discover what effect specific aspects of the surroundings have on the activities essential for life — such as obtaining food and water. It becomes necessary to isolate the separate factors that add up to the total environment. When these are known we can consider the effect of each on the structures and functions of organisms.

Three basic aspects of the physical environment are its surface features and soils, its temperature range, and the amount of precipitation. Combinations of these result in a great variety of habitats, each with its distinctive living conditions. By focusing our attention on the factors present in a specific area, we can arrive at a fair conclusion of the environment surrounding a typical plant and animal. Let us select a spot somewhere in southern New England, in a region of moderate rainfall and having an alternation of a cold winter with a warm summer. Here a common plant is the maple tree, a common animal the chipmunk.

The environment of the maple consists of numerous factors, some present aboveground in the air, others associated with the ground as well as underground. Obvious atmospheric factors are sunlight, air temperature, various gases, and precipitation; ground factors include topography — or surface features — and soil conditions. Sunlight is essential for photosynthesis, the process by which the maple manufactures its food. This is adequate to produce a green plant with as much bulk as a tree. Sunlight also contributes warmth; the temperature range in

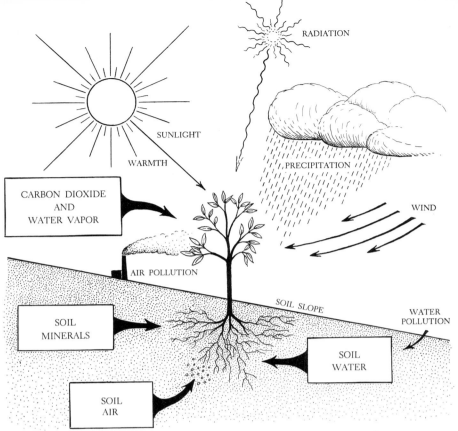

The physical environment of a green plant consists of numerous factors, some present in the atmosphere, some in or on the ground.

this particular area is such that the tree can carry on photosynthesis during enough of the year to maintain its perennial habit of growth. The atmosphere provides enough carbon dioxide so that the maple can secure the carbon needed in food manufacture. It also provides enough water vapor so that humidity is fairly high, preventing excessive water loss by evaporation from the broad leaves of the tree. Precipitation, about forty inches of rain a year, is adequate for tree growth. The rich porous soil can be penetrated by the extensive root system of the maple tree, and provides the minerals and water essential for its activities. Had the ground been deficient in minerals, or made of bare rock, the maple would not be present as a part of the wildlife community. Thus the combination of all these physical conditions explains why this type of plant grows where it does.

The environment of the chipmunk that dashes past the base of the maple consists of the same factors. Some of these are equally important to the animal as to the tree, others are of greater or lesser importance. Sunlight is not as essential since the chipmunk does not rely upon photosynthesis for nourishment. It does, however, provide the illumination regulating its activities from one sunrise to another. It is a daytime-active, or diurnal, animal which searches for its food, becomes aware of its enemies, and carries on other activities dependent upon light. The ground surface and soil are also not as essential, since the chipmunk is free to move about. Nevertheless, the ground does provide an opportunity for shelter and a home in which to raise a family. Temperature is as important to the chipmunk as to the maple because it determines the chipmunk's summer and winter living habits. In this particular region extremes of heat and cold, which might be fatal to the animal, do not occur. Precipitation too is vital to both tree and chipmunk; in this rainfall region it is ample to provide the pools and streams from which the chipmunk can drink. Of the atmospheric gases, carbon dioxide is immaterial unless it accumulates in toxic amounts, but oxygen is essential for the chipmunk's respiration.

The physical environment of a land animal includes many of the same factors as those which affect plants, but some are of lesser significance, some greater.

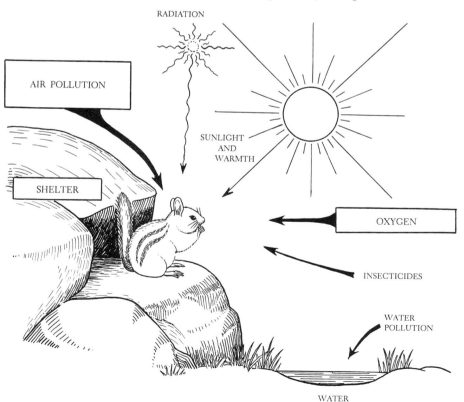

38 / WILDLIFE AND ITS ENVIRONMENT

This is a closeup view of the physical factors in a specific environment. If we had selected another part of the country the same factors would be present. Two of these play a major role in the adaptations of all wildlife, and also in their distribution. They are temperature and precipitation. A third factor is of importance to plants — the nature of the ground surface and soil. Other factors, such as sunlight and atmospheric gases, are so uniform throughout the continent that they are less critical in determining why certain plants and animals live where they do, why they look and act as they do.

How Life Is Adapted to the Temperature Factor

Protoplasm, the living substance of life, is surprisingly adaptable in its ability to endure a wide range of temperatures. Some bacteria are able to survive a low temperature of − 454° F. and a high temperature of 284° F. Killifishes live in water at a temperature of 204° F. and blue-green algae thrive in hot springs at 190° F. At the other extreme, above-ground parts of woody plants can withstand temperatures well below zero. Frogs have survived experimental temperatures of − 18° F. and fishes, − 4° F. The more complex forms of life, found among higher plants and land vertebrates, can only live within a much narrower temperature range, however.

Temperature extremes in North America range approximately from a low of − 70° F. to a high of 130° F., a spread of only 200 degrees. This is well within the limits of life. The temperature at which a particular organism can best carry on its living activities is called the *optimum temperature*. This has a considerably narrower range, and varies not only for different species but for different types of activities. For photosynthesis it is between 60° and 90° F. The optimum temperature for most human functions is relatively narrow also — from 65° to 112° F. When in environments below or above these temperatures, man must protect his body from frostbite or heat prostration. Species whose optimum temperatures for reproduction have a wide range, the coyote, for example, become widely distributed; species such as the manatee have a very limited distribution in North America because of their more restricted optimum temperature.

Plants and Low Temperatures. The effect of low temperatures upon plant life is strikingly demonstrated by the distribution of trees and their type of foliage. Trees require much more water for their growth than do small plants like grasses. When the soil is frozen or when soil water is too cold to be effectively absorbed into the root system, tree growth is reduced or ceases completely. In the Arctic the temperatures are so low for most of the year that the subsoil remains permanently frozen, a condition known as *permafrost*. Only a foot or two of the upper soil thaws in the short summer, preventing the root growth essential for a tree's development. This explains the absence of forests in the tundra and the prevalence of plants such as lichens with no roots at all, or of grasses with shallow root systems. The summits of high mountains have similar low temperatures for most of the year and are also treeless.

Much of North America has a seasonal climate with low temperatures restricted to the winter months but with sufficiently warm and long summer months to thaw the soil completely and provide a growing season long enough for tree growth. In winter the frozen soil reduces the amount of available water for tree roots, and at the same time the air temperature is low, far lower than the optimum for photosynthesis. Water loss from plants is chiefly by evaporation from leaves; such loss of water vapor is known as *transpiration*. With a lessened intake of water, the only way to preserve a normal water balance is to reduce transpiration.

Transpiration is reduced to a minimum in *evergreen trees* by means of leaves that offer a small surface for evaporation of water vapor; these leaves are in the form of needles or scales. To further decrease transpiration, the leaves are coated with an impervious waxy covering. So adapted, evergreen trees can retain their foliage in winter and carry on some photosynthetic activity during warm spells in winter. *Deciduous* trees, such as maples and birches, have become adapted to the same environmental temperature factor by discarding their delicate, vulnerable leaves during the winter. These trees are leafless and dormant during periods of low temperatures, completely eliminating the transpiration problem — but also having to abandon photosynthesis. Deciduous forests develop south of the evergreen forests in the northern lowlands, and below them on mountain slopes.

Plants and High Temperatures. Excessively high temperatures are a hazard to plants only where the water supply becomes limited. High temperatures therefore become a significant environmental factor in arid regions, where the rainfall is less than 10 inches a year.

Temperature and Plant Reproduction. Temperature limits the distribution of many plants by its effect upon reproductive activities. Flowers and fruits cannot tolerate as low temperatures as foliage. It is a familiar experience to gardeners and farmers to see a freezing spell lasting only a few hours destroy all the flowers or fruits. Many plants can be successfully introduced into regions where the temperatures do not hinder vegetative growth but prevent reproduction and seed formation. Date palms are grown as ornamental trees in southern states where they may thrive but do not produce dates. The same is true of the tropical banana plant, which survives in extremely warm parts of the United States but is unable to produce mature fruit.

Animals and Low Temperatures. Temperature is a significant factor in the distribution and habits of animals as well, since it affects the optimum temperatures for their vital activities. This is best seen in the difference between cold-blooded and warm-blooded vertebrates. The cold-blooded amphibians and reptiles are at a great disadvantage when the temperature of the environment falls below the optimum for their bodily activities. These vertebrates are therefore most abundant in warm climates. They decrease in numbers northward on the continent, and at high altitudes. Of the 360 common species of North American reptiles, more than 150 live in Florida, less than 50 in New England, and very few in Canada. Most cold-blooded animals cease to be active at 40° F. As a result, amphibians and reptiles that do live in cold regions become dormant when the temperature drops below this point, retreating beneath the ground or burrowing into the mud at the bottom of lakes.

Birds and mammals, being warm-blooded, are able to maintain their optimum body temperature irrespective of changes in the temperature of the environment. This is an important adaptation, enabling such animals to be active at low temperatures that cause cold-blooded animals to become dormant. The temperature-regulating mechanism consists of a "biological thermostat" which balances the heat lost to the environment by the amount of heat produced in the body. An insulation of fur or feathers aids in conserving body warmth during cold weather, for the optimum body temperature for many mammals is approximately 98° F. Warm-blooded animals are the only animals likely to be found active in the colder northern states in winter.

Low temperature indirectly affects animal life by reducing the available food supply. The scarcity of vegetation — fresh foliage, fruits, seeds — poses a problem for the herbivores. As the herbivores become less active and available, the carnivores find it more difficult to find food. Many mammals and birds remain active the year round, as for example the lemmings, hares, foxes, and wolves. Others adapt to the diminished food supply in one of three ways: by migration, by becoming dormant, and by hibernation. In the western mountains, elk have summer and winter feeding grounds, migrating at the approach of winter from the higher summer pastures to slopes at lower elevations where more vegetation is available. Although not a primary cause of migration in birds, reduced food supply undoubtedly plays a role in their flight from their summer breeding grounds in the Arctic to regions with milder temperatures and more food.

Mammals like chipmunks and bears adapt to food scarcity by taking a long winter's sleep. In the dormant condition their bodies use a minimum of food reserve, consuming the fat stored in their bodies until spring. With certain mammals the dormant period is interrupted by short active periods enabling the animal to go to its cache of food and have a midwinter snack. Others exhibit a unique adaptation known as *hibernation*, an adaptation developed by the woodchuck and ground squirrel. Low temperature initiates the hibernation, which resembles deep sleep. During hibernation, the "thermostat" in the brain which regulates body temperature is reset from the normal 98° F. to a lower temperature, in the 40°–50° F. range. At the same time the heartbeat of the hibernator slows from a possible 200 per minute to only 5 or 10, and the respiratory rate is so drastically reduced that oxygen consumption is 7 per cent of normal. These modifications of basic functions permit the animal to remain alive during the winter months, using very little reserve food.

Animals and High Temperatures. High temperatures also may present a problem to animals, especially the cold-blooded ones. Reptiles are popularly believed capable of enduring extremely high temperatures since they are so common in arid habitats. On the contrary, most reptiles cannot endure as much heat as mammals. The optimum body temperature for many reptiles is about 104° F. Lizards die when their body temperature reaches 113° F.; yet desert temperatures in the sun,

The crested lizard in summer retreats to its burrow to escape the high noonday temperatures of the desert.

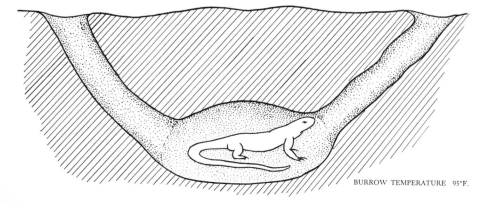

during summer, often exceed 104° F. Therefore most reptiles at such times make only quick forays into the open for food and soon dart back into the cool shade of vegetation or rocks. Desert animals are most active in early morning and late evening. Many reptiles retreat to burrows to escape the heat and desiccation; the gopher tortoise retires to a special summer burrow during midday; and lizards often have air-conditioned underground homes where the temperature is a cool 95° F. while the surface sand reaches a sizzling 120° F. Some desert animals pass excessively hot periods in an inactive type of dormancy known as *estivation*. Others — the desert bird known as the roadrunner is an example — seem oblivious to the heat and are active even on the hottest days.

How Life Is Adapted to the Water Factor

A major problem facing land-dwelling plants and animals is how to obtain sufficient water and, once having secured it, how to conserve it. Since the ultimate source of all water on land is rainfall, this factor in the environment is a very critical one. Precipitation, as we have seen, varies from great scarcity to a superabundance and so creates habitats varying widely in the availability of water. Water is a factor in the environment while yet a vapor in the atmosphere, when falling

Lichens are cold-climate thallus plants that can live without soil.

as snow, hail, or rain, and when it finally accumulates in the ground. In all three situations it plays a role in the distribution of wildlife and its adaptations to meet too little or too great amounts of water.

Humidity and Wildlife. Humidity, caused by the presence of water vapor in the atmosphere, determines the amount of moisture available directly from the air. Some plants can get their water from this source, and hence can do without roots. Lichens are plants with this ability; they are abundant in cool regions where the humidity is high. Accordingly, lichens are common in the Arctic and in the cool humid forests of the North. Other air plants are typical of warm humid regions. Some, attached to the larger plants as trees for support, are called *epiphytes*. Common epiphytes are Spanish moss and orchids.

Humidity is also important to land plants and animals because of its effect on evaporation of water from their bodies. When humidity is great, water loss by transpiration is reduced to a minimum. This makes it possible for plants, even when they lack special adaptations for reducing transpiration, to live in humid habitats. Mosses and ferns, whose delicate leaves are sensitive to drying out in direct sunlight, are for this reason abundant on the floor of dense forests in regions of great rainfall, as in the Olympic Peninsula. Amphibians are animals with thin skins which must be kept continuously moist. Thus they are poorly adapted to

Common epiphyte of warm humid regions in the southeastern United States is the leafless Spanish moss.

A hydrophytic community consists of plants adapted to live in very wet habitats, such as this swampy pond in Acadia National Park, Maine.

live in an environment where low humidity would cause fatal drying out of their body surface. For this reason such amphibians as frogs and salamanders reach their greatest abundance in warm and humid forests. Water in the air, as fog, also reduces transpiration and can be a factor in the distribution of trees. Maximum tree growth occurs in the fog belt on the seaward side of the coastal mountains of Oregon and California, as evidenced by the great redwood forests.

Water and Plant Life. The major role of precipitation, however, lies in its accumulation as water on the surface of the land or stored in the ground. On the basis of their water needs, and their adaptations to varying amounts of water in the environment, plants can be grouped into three types: hydrophytes, mesophytes, and xerophytes. *Hydrophytes* (from the Greek words *hydros* meaning "water," and *phyton*, meaning "plant") are plants adapted for living in very wet habitats like ponds, streams, and swamps. They are found in regions of ample rainfall — 40 inches or more annually. Consequently they are abundant in the eastern and southern United States where precipitation is high and topography favors an accumulation of soil water. An excess of water in the soil is a handicap to many plants because it reduces the oxygen content of the soil air spaces and so interferes with respiration of roots and

Bald cypress is a hydrophyte with special aerial roots known as knees, which aid in bringing air to the root system.

developing seedlings. Roots of many hydrophytes are less sensitive to this oxygen deficiency; seedlings of a hydrophyte such as cattail can germinate in a swamp because of this tolerance to the low oxygen content of water-logged soil. Larch and red maple are common hydrophytic trees, but most completely adapted to wet ground are the bald cypresses. Bald cypress can grow even when its roots are covered by several feet of water. The root system produces aerial projections known as "knees," with thin bark and a hollow center. Rising above the water level, these knees become breathing tubes through which air can reach the root system. Plants that spend their entire lives in the water are the most complete hydrophytes. Such is the water hyacinth, a weed of southern streams and lakes whose floating leaf clusters have stems with bladder-like enlargements. These are not only reservoirs of air but also a flotation device, freeing the plants from the necessity of being rooted in the soil.

Mesophytes (from the Greek *mesos*, meaning "in the middle") are plants living in habitats where the water supply is moderate — neither very abundant nor very scarce. They are common in areas which have an annual rainfall of from twenty to forty inches. Mesophytes include the great majority of our trees, shrubs, grasses, and smaller flowering

A mesophytic community, such as this deciduous woods in Hoosier National Forest, Indiana, lives in habitats where the water supply is moderate, neither very abundant or very scarce. (*Credit: U.S.D.A.*)

plants. The sugar maple we met earlier in this chapter is a typical mesophyte, as were most of the plants we discovered on our first ecological adventure. Such plants, as might be expected, exhibit few special adaptations for the simple reason that they live in an environment where there is neither too much nor too little water.

Xerophytes (from the Greek word *xeros,* meaning "dry") are plants adapted to live where water is scarce — in deserts or on windswept mountains. They are the dominant plants in areas with less than ten inches of rain a year, or in areas where high winds increase the transpiration rate. Xerophytes form the characteristic vegetation of deserts and exhibit many unusual adaptations that aid these plants to survive in arid habitats. The adaptations are of three types: those which lessen the loss of water by transpiration, those which conserve water in storage tissues, and those which make the root system more effective in obtaining water. *Reduced transpiration* is brought about, as in evergreen trees, by development of small leaves with a waxy or hairy surface; this method of lessening transpiration can be seen in the foliage of sagebrush and the creosote bush, two common desert shrubs. Transpiration is also reduced by discarding the leaves during dry spells. The cacti are the champion water conservers among all xerophytes. Cacti have no leaves at all; it is the green compact stem that assumes the task of photosynthesis. Transpiration of the stem is reduced by an impervious outer surface and by

A xerophytic community consists of plants adapted to live in very dry habitats. Cactus community on a southern Arizona hillside.

the interlacing spines, which create a blanket of air close to the stem, more humid than the air of the environment.

Water conservation by storage is accomplished by special types of leaves, stems, and roots. Storage of water in the underground parts of the plant is to be found in an Arizona night-blooming cactus, although among most of the cacti water is stored in the enlarged and succulent stem. The numerous kinds of barrel and hedgehog cacti illustrate this principle. When rain does come to the desert, plants must be able to absorb as much water as possible, often as rapidly as possible. Thus *water absorption* becomes another important factor in the xerophyte's life. Successful xerophytes obtain their water by two types of root systems, one of which relies on extending as far as possible in all directions around the plant, the other, on penetrating deeply into the subsoil.

Xerophytes have a special problem when it comes to reproduction, because flower production and seed germination must coincide with the rainy season. Many desert annuals bloom at the same time, immediately after the first spring rains, covering the desert floor with a carpet of colorful hues. Seeds are produced rapidly, often within a few weeks. But if the seeds germinated at once they would perish in the ensuing dry season. It may also be months, even years, before another rainy season makes seed germination possible. Many seeds are adapted for such an emergency by a coating that prevents immediate germination. Consequently they lie dormant until a wet spell provides sufficient moisture to dissolve the resistant outer covering of the seed. The seeds of some xerophytic annuals require a period of several years before the seed coat can be effectively dissolved for germination to take place.

Water and Animal Life. In regions of low humidity and water scarcity, animals face the same water problems as those confronting plants. They adjust to this in two ways: by adaptations that conserve their body water content and by obtaining water in other ways than by drinking. Reptiles are more common than amphibians in arid environments because their scaly body covering reduces water loss by evaporation, sealing in their body moisture. Most reptiles, as we have seen, avoid extreme summer heat by retiring to their burrows and so reduce

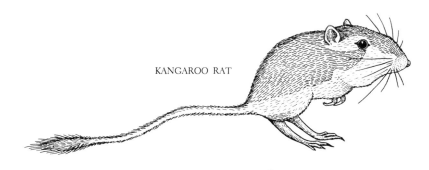

KANGAROO RAT

further any chance of excessive water loss. Reptiles also are independent of water during their reproductive phase, since their eggs develop on land while buried in the sand.

Both reptiles and desert mammals such as the herbivorous pack rats and jack rabbits obtain much of their water by eating succulent vegetation, and can therefore go long periods without drinking.

The most extraordinary adaptation, however, is found in the horned lizards and the kangaroo rat, which have the ability to manufacture their own water. The friendly kangaroo rat of the southwestern desert is no relation to the kangaroo but gets its name from the long hind legs and jumping method of locomotion. Known as the rodent that never drinks, it has been called "the most triumphant imaginable example of adaptation to the most characteristic desert difficulty, the absence of water." Water is a simple compound, made up of the two elements hydrogen and oxygen. Green plants, during photosynthesis, break down water into its two constituent elements and recombine them as carbohydrates. Nevertheless, they cannot reverse the reaction and make water out of the oxygen and hydrogen found in carbohydrates. Yet the kangaroo rat does exactly this. Its food consists of dry seeds, which contain a considerable amount of carbohydrates. From these the kangaroo rat can extract oxygen and hydrogen and combine them to form *metabolic water*. These desert rodents even refuse water when it is offered them. Thus their herbivorous diet furnishes them both with food and with drink. This unique ability gives its possessor a great advantage over other animals in a region where it rarely rains.

Life and Light

The important portion of the light spectrum, as far as life is concerned, are the wavelengths between the infrared and the ultraviolet extremes. These are of fairly uniform distribution over the entire continent. As a result this aspect of solar radiation is not a critical factor in determining adaptations and distribution of organisms. Light does nonetheless vary in intensity and in duration, each of which affects the lives of plants and animals.

Many plants, especially trees, compete with each other to reach open areas where light intensity is at its maximum. Below the forest canopy the light intensity diminishes, and the ground is often bare of vegetation or may support only those plants capable of carrying on photosynthesis in dim light. The ability to grow in light of low intensity is known as *tolerance*. Tolerant plants are shade plants; they form the thick carpet over earth, fallen trees, and boulders on the forest floor. Many are mosses and ferns, but some are small flowering plants.

Light intensity is also important to animals, but to a lesser degree. Some animals are diurnal, active by day and so dependent upon light for their food. Others are nocturnal, possessing special adaptations that enable them to find their way about in the dark. Owls are bird carnivores adapted through their special type of vision for seeing in the dark, whereas those mammal carnivores, the bats, find their way in the dark by a remarkable sonar system.

Life and Other Physical Factors

Wind is an important factor in the environment of plants, largely because of its effect on increasing transpiration. Winds also affect the growth of trees, limiting the growth of branches to the lee side of the main trunk, and forcing the trees into a prostrate and spreading position. High winds are typical of mountains, where the continuous gales prohibit life of any kind on exposed ridges. The highest wind velocity ever recorded was on a mountaintop: 234 miles per hour on the summit of Mount Washington in New Hampshire. It is a wonder that any vegetation can survive when battered by such tremendous force.

Winds also cause a shifting of the soil, if it is made up of sand and loose particles of rock. Seedlings have difficulty in establishing them-

Wind determines the shape and branching of timberline trees, such as this limber pine in Wasatch National Forest, Utah. (*Credit: U.S. Forest Service*)

selves in the unstable substratum, and only those plants survive which can quickly anchor themselves with a spreading and tenacious root system. Many sand dunes of western arid regions are populated only by grasses and deep-rooted yuccas.

The nature of the earth's surface plays a more important role in the lives of attached organisms, such as plants, than in those of the mobile animals. Some soils have an excess of minerals that make the environment of plant roots very alkaline; other soils have accumulations of organic material which result in an acid soil. In either case, only those plants with an ability to grow under such conditions can survive. The desert saltbush, found on alkali flats of the arid western plateaus, and the succulent glasswort of sea beaches are plants which can live in saline soils. In the acid soils of bogs the dominant plants include the insect-eating pitcher plants and the larches. Within every forest area, local conditions causing such variation in soil acidity or alkalinity create special environments in which certain species of trees can be found.

The degree of the slope, and its exposure, is also a factor of some importance in determining the kinds of plants which can grow in an area. North-facing slopes are cooler and moister than those facing south, and therefore often have a different type of plant community. This in turn affects the distribution of the animals dependent upon the vegetation for food or shelter. On the north-facing slope of a hillside

The north-facing and south-facing slopes of a ravine present environments for different communities of plants.

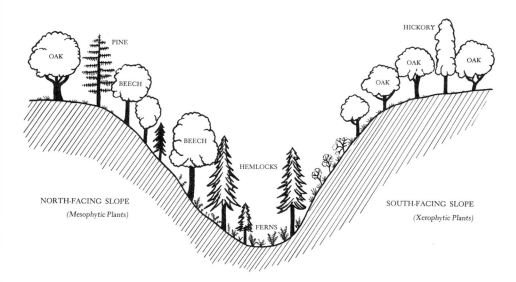

in the eastern United States, beeches, hemlocks, ferns, and mosses will thrive. On the drier south-facing slopes, oaks and hickories grow among grasses and xerophytic wildflowers. In the Rocky Mountains, north-facing slopes are covered with Douglasfir; the valley bottoms are forested with blue spruce; and ponderosa pine thrives on the south-facing slopes. In the Sierra Nevada, the western slopes have abundant water and support thick forests of redwoods and sugar pine; the arid eastern slopes, especially at lower elevations, are covered with piñon pine and juniper.

The surface of the ground, closely related to the topography, affects the distribution of animals by presenting different problems in locomotion. In bare rocky areas the mountain goat has an advantage over the clawed mammals which may be pursuing it. The speedy hoofed antelope excels on the flat prairies, where it can develop the speed needed to escape its enemies. In sandy areas the burrowing prairie dogs can build their underground homes, a difficult feat if they lived on a rocky New England hillside.

This completes the story of the ways in which plants and animals have become adapted to living with the physical factors presented by the varied topography and climate. But sunlight and wind, rain and snow, warmth and cold are only part of the story. There remains the living environment — the many different kinds of plants and animals which occupy the same habitat. To appreciate the environment in its entirety, we must also see how living things affect each other, and how these interrelations create a wildlife community.

4 / Adaptation of Life to Its Living Environment

Every organism must be adapted not only to its physical environment but, of equal importance, to its living surroundings. Whether plant or animal, it must make adjustments to living among other organisms. The many ways in which plants and animals are adapted to living with each other results in a complex network of interrelations aptly called "the web of life."

In a wildlife community, as in a human one, the main business of life is obtaining food. Members of the community use different methods for securing their food, depending upon whether they are green plants, nongreen plants, or animals. Often these methods involve a dependence upon one another which may prove beneficial or harmful to other organisms. In addition to obtaining nourishment, each member of the community must assist in perpetuating its species. To accomplish this, plants and animals possess other adaptations that may also involve dependence upon one another. Whether for self-maintenance or for self-perpetuation, the relations of living things to each other are varied and complex.

Green Plants and Decay Bacteria

Food is the source of all the energy which makes a wildlife community a living one. The first step in the production of food is the transformation of the radiant energy in sunlight into the potential energy of carbohydrates (starches and sugars). Additional steps in food manufacture result in the synthesis of fats and proteins. All of these basic foods are composed of carbon; proteins in addition contain nitrogen. Thus carbon and nitrogen are key elements in food production. They occur in many forms, but only two of them can be used by green plants in food manufacture. Carbon dioxide of the atmosphere is the source of the carbon and soil nitrates are the source of the nitrogen. Green plants are continually withdrawing carbon dioxide from the atmosphere in the process of manufacturing carbohydrates. During protein synthesis they are removing nitrates from the soil. The supply of these compounds is not

54 / WILDLIFE AND ITS ENVIRONMENT

unlimited; yet life, with its dependence upon food, has been consuming them for more than a billion years. Why has the supply of carbon and nitrogen not been used up? The answer lies in the special abilities of some bacteria to renew the store of carbon dioxide and nitrates, making it possible for plants to use the carbon and nitrogen in an endless cycle. The replenishment of the carbon takes place in the carbon cycle, of nitrogen in a nitrogen cycle.

Carbon Cycle. Decay is universal but its significance is often overlooked. We associate it with the destruction of food, with a rotting log or a decaying carcass. To many of us the word may have an unpleasant connotation. Nevertheless, decay is essential for the basic economy that pervades all nature. It is the destruction of muscle and bone, of wood and foliage through the action of those ubiquitous nongreen plants the bacteria. The earth would be a sorry sight indeed if its surface were heaped with the remains of all the organisms that had ever lived. Decay

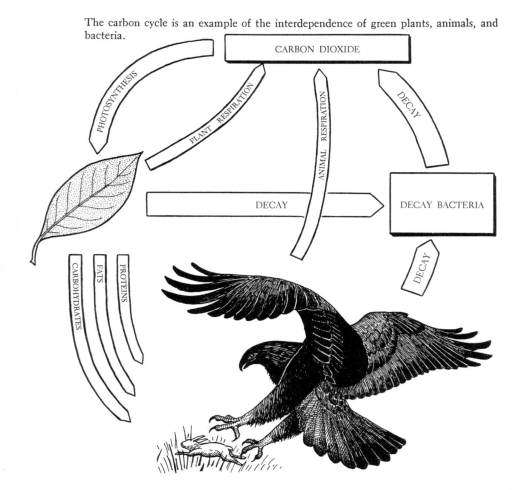

The carbon cycle is an example of the interdependence of green plants, animals, and bacteria.

bacteria perform the valuable community service of removing such vestiges of the past. Yet far more important is the chemical transformation they bring about whereby complex protoplasmic substances are reduced to their simpler components — one of which is carbon dioxide. By this process, the carbohydrates, proteins, and fats are chemically changed into the simpler compounds out of which they were made. In this way carbon is returned to the atmosphere, from which it previously has been extracted.

Green plants convert most of the carbon they have taken from the air into the tissues of their bodies: into roots, foliage, branches, fruits, and seeds. When plant tissue is eaten by an animal, the carbon becomes part of the compounds making up the animal's tissues. As generation after generation of plants and animals increase their bulk by growth, more and more of the carbon is removed from the air. A very small percentage of it has been returned to the atmosphere through animal respiration and combustion. Neither of these two processes would contribute enough carbon dioxide to long postpone the fateful day when all the carbon dioxide was completely absorbed from the atmosphere and all vegetation would disappear from the earth. Only 0.035 per cent of our atmosphere is carbon dioxide, an amount that could support the present plant population for less than forty years! Therefore the role of decay bacteria in returning the carbon to the atmosphere is a critical one. Otherwise, increasing amounts of carbon would be locked up forever in the remains of life. This vital interrelation of green plants and bacteria results from the fact that decay is the way of life by which these simple organisms obtain their food from the organic material in dead plants and animals. In the community bank, carbon is continuously being withdrawn by the green plants, but decay bacteria are at the same time depositing funds to prevent a deficit.

Nitrogen Cycle. Four fifths of the atmosphere consists of nitrogen; surprisingly, few green plants can make use of this ample supply. The nitrogen used in protein synthesis comes from nitrates — nitrogen compounds found in the soil and absorbed by the plant's root system. As in the case of carbon, the nitrate supply is not unlimited. Extraction of the nitrate reserve in the ground would eventually use up this source of nitrogen. Through growth and death, nitrogen becomes locked up in the protein compounds of plant and animal bodies. Here, too, bacteria play an important part. Some bacteria, in the course of their feeding habits, are able to break down proteins into ammonia, a simpler nitrogen compound. Other bacteria can change the ammonia into the still simpler nitrites and the nitrites in turn to nitrates. Thus again, in the

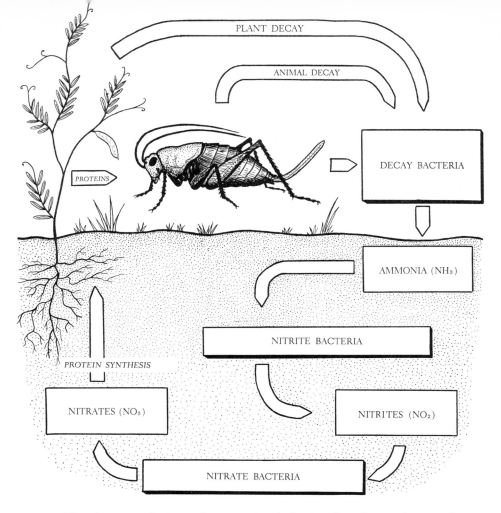

The nitrogen cycle is another example of the interdependence of green plants, animals, and nongreen plants in the community.

community bank, the bacteria make the deposits to the nitrogen account while the green plants are constantly making withdrawals.

Have you ever asked yourself why one so seldom sees traces of any dead animal or plant in the woods? One reason this is true is the effectiveness of the decay organisms that undertake the double task of sanitation engineer and bank depositor of the vital carbon and nitrogen. If you are disturbed by the sight of a rotting, fallen forest giant or the decomposing body of a squirrel, remember that decay is necessary for the never-ending cycle of elements upon which continuous rebirth of life depends.

Green Plants and Herbivores

The food manufactured by green plants is used primarily by themselves, but so efficient are the photosynthetic and protein-building processes that an excess of food often results. Foliage is produced in far greater quantity than the plant requires. Reproductive organs like fruits and seeds are produced in amounts far beyond those needed for perpetuation of the species. Herbivorous animals profit by this excess food production; their consumption of plant tissue usually does no harm to the plants themselves. Herbivores have developed special adaptations for a herbivorous diet, and in some instances plants have reciprocated by developing adaptations that protect them from complete destruction by hungry herbivores.

Adaptations of Herbivores. Insects are the most numerous of all herbivores, making up for their small size by their voracious appetites. As any gardener or farmer well knows, every part of a plant provides food for some kind of insect. In the seed may lie eggs and larvae. The newly emerged seedling falls prey to grubs, cutworms, and other soil-inhabiting insects with a fondness for tender plant tissues. Leaves are consumed by locusts, woody stems by beetles, fruits by the larvae of moths. A single white pine may provide sustenance and home for more than two dozen species of plant-feeding insects, and more than 1000 different insects feed on oak trees. The mouthparts of an insect are specialized in many ways for securing plant food. In the chewing insects — the grasshoppers, locusts, and leaf beetles — the mouth appendages form jaws that operate sidewise like scissors and are equally effective in cutting up leaves. The larvae, or caterpillars, of butterflies and moths also have chewing mouth parts and a notorious liking for foliage; best known is the tent caterpillar. In the sucking insects the mouth appendages form hypodermic needles which can puncture plant tissues and siphon off the nutritive plant sap. Plant bugs and aphids are widespread sucking insects. Adult moths and butterflies have a specialized proboscis enabling them to suck nectar from flowers. The list of insects feeding on plants is a long one indeed, indicating the success with which the insect world has put the plant world to its own use.

Many birds are also herbivorous, but since they prefer fruits and seeds to foliage they are not as great a threat to plant life. Birds like goldfinches and sparrows have stout conical beaks capable of cracking open nuts and hard seeds. Game birds, in addition to having beaks of the same type, possess clawed toes used in scratching for fallen fruits and seeds. Because they also devour vast quantities of insects, birds prevent the excessive damage to plants which otherwise might take place. Were it not for the birds, green plants might find it difficult to survive the onslaughts of the hordes of herbivorous insects which frequently occur. And so we see another strand in the complex web which entangles the members of a wildlife community.

Of the herbivorous mammals, rodents and ungulates reveal numerous adaptations for eating plant food. The chisel-like teeth of rodents enable them to cut down trees, as the aspens felled by beavers testify. Foliage is often coarse food, needing to be ground into small pieces before it can be digested. The broad grinding molar teeth of ungulates such as deer and moose are suited for grazing and browsing. Gophers and other burrowing mammals feed on roots and tubers. Fruits and berries are eaten by many mammals otherwise carnivorous; foxes and bears are particularly fond of such a diet. Nuts and acorns are a staple diet of squirrels and chipmunks, often being collected into hidden stores to be used as a winter food reserve.

Plant Protection against Herbivores. Green plants are able to survive the onslaught of herbivores in many different ways. When fruits and seeds are eaten, the individual plant does not suffer since these parts play no part in maintaining the life of the plant itself. When foliage is eaten by mammals, many of the leaves may be out of reach so that sufficient foliage remains to carry on adequate photosynthesis. Even when complete defoliation occurs, as may happen after an attack by locusts, a new crop of buds often saves the plant's life. Grass leaves grow from the base, so that the tips can be eaten without damage to the plant. Some plants have special adaptations to protect them from herbivores. Many herbs and shrubs possess thorns or spiny foliage and unpleasant-tasting or poisonous substances which repel animals. In arid regions, where vegetation is scarce, an unprotected succulent plant is greedily eaten by the first passing herbivore. Desert plants therefore are conspicuously thorny, as anyone who has encountered mesquite, acacia, or cacti will agree. Possession of spines as a defensive armor is an obvious advantage to a green plant anywhere, but in a desert community spines become a necessity for survival.

Survival by Cooperation

Nature is often considered a battleground where aggressive behavior is essential for survival in competition with one's neighbors. This idea is implied in the familiar expression "struggle for existence." The competitive relationship does not always hold true, however, for plants and animals have found that cooperation can also be of use in survival. A green plant, for example, possesses the special ability to manufacture its own food, an ability an animal lacks. A land animal, on the other hand, can move about, something impossible for a plant to do. It is not, then, too surprising that some plants and animals have solved the problems peculiar to their way of life by using the special ability of the other. We have seen how herbivores are dependent upon green plants for food. This may seem a one-sided arrangement until we find that in exchange plants, in solving their problems of seed formation and dispersal, make good use of the animal's ability to move.

Animal Cooperation in Plant Reproduction. The formation of seeds in both gymnosperms and angiosperms is dependent upon transport of the pollen from the stamens, where it is produced, to the ovules, or plant eggs, located at the base of the pistil. Plants growing close together can rely upon air currents to carry pollen from one flower to the other. This takes place in conifers, catkin-trees, and grasses. But for plants that grow in isolated groups — many herbaceous and woody species are examples — wind is not a reliable transporting agent. Many millions of years ago seed plants found an alternative method: the use of animals to carry pollen from flower to flower. Such plants advertise for carriers by developing brightly colored petals or adding sweet nectar as a special bribe. Bees and wasps are the most frequent pollinating agents of flowering plants. Often the flowers of one species are adapted for pollination by only one species of insect; the mouthparts of this species are likewise suited for only that particular flower. The resulting interdependence of flowers and insects is an outstanding example of plant-animal cooperation.

Animal Cooperation in Seed Dispersal. Another problem confronting an immobilized plant is the dispersal of its young. The embryonic plant is encased in the seed and remains there until the seed germinates. If all the seeds produced by a plant were to drop to the ground and germinate near the parent, only a few seedlings would survive in the overcrowded environment. Survival of a species depends upon dispersal of a new generation to another suitable habitat. Again, many seed plants make use of the wind for such dispersal as well as for pollination. These plants have fruits and seeds equipped with wings or parachutes, as in the cottonwoods and maples. But other plants make use of animals to disperse their seeds. Animals transport the seed a much greater distance; they also can carry a larger seed within which more food for the developing seedling can be stored. Here again some plants offer bribes in exchange for the animal's cooperation. Many fruits are brightly colored to attract the animal's attention, and contain edible juicy contents. After the fruit is eaten, the seeds pass unharmed through the animal's digestive tract and may be deposited many miles from where they were eaten. Another device is the development of fruits with spiny or prickly jackets that cling to the animal's body. Such burs and stick-tights often are carried a great distance before they are shaken off. This method of plant dispersal is a kind of hitchhiking that enables otherwise immobile organisms to travel.

Other cooperative methods involve a closer relation between two species; this is known as *symbiosis*. The term is based on two Greek words meaning "living together." Symbiosis usually is applied solely to those living arrangements beneficial to one or both partners, and not harming either. When only one partner benefits from the symbiosis it is known as commensalism. When both partners profit by the association it is called mutualism.

Commensalism (from the Latin, meaning "sharing the same table") is a type of symbiosis more common among animals than plants. The most familiar example is that which exists between the remora, or sharksucker, and a larger fish. The remora is an animal hitchhiker that manages a free ride while attached to a large fish such as a shark; at the same time, it secures free meals from the remnants of food the larger carnivore leaves when devouring its prey. Another type of commensalism is that of cattle egrets and cattle. The egrets congregate around the grazing animals and feed on the swarms of insects stirred up in the grass when the larger animals move about. This furnishes the egrets with ample food; such a relation neither harms nor benefits the cattle. Plant epiphytes can be considered a type of commensalism. Spanish moss, which attaches itself to trees for support, exemplies this. Such plants do not harm their partners unless present in enough quantity to put excessive strain on the branches.

Mutualism is a more widespread type of symbiosis, with many examples among green plants, nongreen plants, and animals. The interdependence of yuccas and a certain pollinating moth is a remarkable example of mutualism involving a green plant and an animal. The yucca, that successful xerophyte of the southwestern United States, bears at its summit a large cluster of lilylike flowers that open at night. Hidden among the flowers are tiny silvery-white moths whose reproductive activities are closely bound up with seed formation in the yucca. The moth collects pollen from the stamens of one flower, then moves on to another, where it pries deep into the blossom to reach the ovary at the base of the pistil. The moth punctures the ovary with its sharply pointed egg-laying

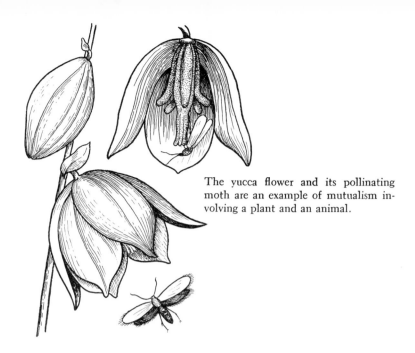

The yucca flower and its pollinating moth are an example of mutualism involving a plant and an animal.

organ and forces its eggs into the ovary tissue. Upon leaving the flower the moth deposits the pollen on the stamens. In due time two events take place. The pollen fertilizes the yucca ovules, this resulting in formation of seeds and a podlike fruit. At the same time the moth eggs laid in the ovary tissue become larvae, whose need for food is satisfied by eating the nearby yucca seeds. Since the plant produces several hundred seeds, and since there are rarely more than four moth larvae to feed upon them, enough seeds are left to produce new yucca plants. When mature, the larvae bore their way out of the pod, fall to the ground, and make a cocoon. Eventually a new moth emerges and the process is repeated. A remarkable feature of this precisely timed procedure is that the moth does not take any pollen or nectar for its own use. What better cooperation could exist between two organisms in solving their reproductive problems? This intimate relationship does present the danger that always accompanies excessive specialization: should either partner in the symbiosis disappear from the community, it would bring extinction to the other. Both the moth and the plant are so highly adapted for each other's services that no other insect can pollinate a yucca flower, and the moth does not lay its eggs in any other plant.

Another type of mutualism is exhibited by lichens. These strange plants can live where no other organisms can survive; one reason for this is their unusual makeup. The bulk of the lichen's body consists of fungus threads that form a tangled network amid which live green algae. Like all green plants, the algae carry on photosynthesis, using atmospheric carbon dioxide and moisture; the fungus threads penetrate the algal cells and absorb food from them. The fungi contribute acids that disintegrate rock, and anchor the lichen to the substratum. They also form a water-conserving shield around the delicate algae, which otherwise could not live on land. In this way the life of a lichen is a constant give-and-take

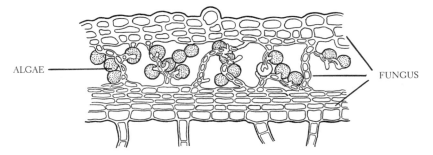

A lichen plant is an example of mutualism involving a green plant and a fungus.

between the green and the nongreen partners which results in a type of plant that can live in barren habitats, independent of soil and soil water.

In a mutualism existing between green plants and bacteria, the roots of legumes such as clover or alfalfa possess nodules of nitrogen-fixing bacteria. These bacteria are capable of extracting nitrogen from the air found in the soil and converting it into nitrates. Nitrates are essential for protein synthesis by the legumes. The root bacteria find lodging in the plant tissue, which provides them with a source of organic food. The plant, in exchange, obtains an extra supply of nitrates. These bacteria have become so dependent upon their green partners that when isolated from them they cannot carry on the nitrogen-fixation process. Because of this type of mutualism, legume crops enrich the soil when they are plowed under at the end of a growing season.

A type of mutualism involving two different kinds of animals is illustrated by termites and protozoa. Termites are insects with the unusual ability to eat wood — ordinarily an indigestible food for animals. Termites are able to subsist on wood because they harbor certain species of protozoa in their digestive tract capable of digesting cellulose, the chief component of wood. These microscopic partners thrive on the wood the termite eats, and the termite in turn makes use of the more digestible foods synthesized in the protozoan body. Experiments reveal that when their protozoa partners are removed, termites starve to death because they can no longer make use of the wood they normally eat.

Survival by Parasitism

It is only a short step from symbiosis to an association of two organisms in which one takes advantage of the other. The plant or animal benefiting by the arrangement is the *parasite*, the one that is exploited is the *host*. The relation between the two species is an unequal partnership in which the parasite has all the advantages and the host is often injured or even destroyed by the unbidden guest. Parasitism has been adopted as a way of life by many thousands of species of plants and animals. It provides a method of obtaining sustenance the easy way — at the expense of another living organism. The adaptations of parasites to accomplish this end are often incredibly complex and, to us, even diabolical. Nature, however, is more concerned with the survival of the species than with the fate of the individual. There is no doubt that the parasitic technique for survival is very successful. Parasitism can involve, either as host or parasite, a green plant, a nongreen plant, or an animal.

Least common is the association of a *green plant as parasite and another green plant as host*. This is the situation in the species of mistletoe used as Christmas decoration. Throughout the southern and western United States mistletoe plants occur as bunches of foliage on the branches of oaks and sycamores. The evergreen mistletoe is conspicuous in winter when the deciduous host is leafless. Since mistletoe is green it can carry on photosynthesis, but it obtains its water and minerals from the tree. It does this by special roots that grow into the branches of the host and penetrate the conducting channels of the wood. The parasite may cause deformed branches and "witches brooms," although usually the host tree is not seriously handicapped by the parasite.

As might be expected, parasitism is very prevalent among the fungi and bacteria because of their inability to make their own food. Some are parasites of green plants, others of animals. The association of *nongreen plants as parasites and green plants as hosts* is the cause of many plant diseases which often eliminate a green plant from the wildlife community. In one type of parasitism the nongreen organisms are bacteria. These microorganisms invade the bodies of plants through the breathing pores of leaves, through cuts and wounds in the bark, or by penetration of the soft tissues of stems and roots. Some bacteria remain in the leaves, causing discoloration and the deformed foliage typical of many plant diseases. Others block the water-conducting channels; they interfere with the movement of water within the plant and thus cause wilting of the foliage. Still other kinds of bacteria stimulate host cells to abnormal growth, causing galls and tumors. In all cases, the methods by which the bacteria obtain their food bring about abnormal conditions in the host tissues and resulting diseases.

One can see a vivid demonstration of such parasitism taking place today, in the desert community of which the saguaro cactus is a member. This well-adapted xerophyte has won a victory over the physical environment, but is threatened with extinction by becoming host to a bacterial parasite. The huge cacti are helpless in the face of an invasion by bacteria carried in the eggs and larvae of a small moth. The moth excavates tunnels in the succulent saguaro stem, contaminating it with its parasites. Small circular spots appear on the trunk and soon enlarge to form purplish, oozing areas. As the disease progresses the base of the saguaro decomposes into a rotting mass, and the desert giant falls to the earth. Destruction of all infected plants is the most effective control of the spread of the disease.

In another type of parasitism the nongreen organisms are fungi. These include a great many molds, mildews, rusts, and rots which infest the photosynthetic members of the community. Wherever there is a green plant, dozens or even hundreds of fungus species lie in wait to parasitize it. Every air current scatters millions of the microscopic spores that, once inside the plant, germinate rapidly into a mass of colorless threads known as a mycelium. Mycelium is to a fungus what roots, stems, and leaves are to a seed plant; it is the maintenance system of the fungus. The mycelium spreads rapidly between the host cells, puncturing them by suckerlike processes which absorb nourishment from the living cells. The infected organ is sometimes so damaged that the injury is permanent and destroys the plant. Eventually the fungus produces spores on the surface of its host, which may be brightly colored red, orange, or yellow.

The most dramatic example of the destructiveness of a fungus parasite is the extinction of the American chestnut forests, once abundant in the eastern United States. The chestnut became host to the chestnut-blight fungus, introduced into the United States on nursery trees brought from the Orient to New York City in 1904. The blight spread rapidly, in areas from Vermont to Pennsylvania; by 1911 thousands of acres of chestnut forest had been wiped out. The spores of the parasite gain entrance through bark wounds and germinate in the underlying living tissue. As the mycelium spreads through the conducting channels of the tree, food and water supplies are cut off and death of the tree is only a matter of time. The first symptom of the disease is a yellowing of the foliage; then the twigs and branches become discolored and a reddish canker appears on the bark. The canker area produces spores that spread to new chestnut trees. Should you travel along the Blue Ridge Parkway through the southern Appalachians you will see many miles of ghost forest with erect but dead weatherbeaten trunks of chestnut trees. They serve as a grim reminder of the losing battle one member of a wildlife community waged against an unchecked parasite.

In parasitism involving *nongreen plants as parasites and animals as hosts,* bacteria rather than fungi are the parasites. The role of bacteria as disease parasites in warm-blooded animals is familiar and are as widespread among wild animals as among domesticated animals and man. Such bacteria usually enter the host's body in food and water, through wounds, and carried by insects. Brucellosis is a bacterial disease of cattle and related ungulates which causes undulant fever when transmitted to man. Other bacterial diseases include tularemia of squirrels and rabbits, bubonic plague of rodents, anthrax of cattle and such carnivores as mink and fox. Many of these diseases are carried from one host to another by ticks, lice, flies, and mosquitoes. Such bacterial diseases often account for the unexpected reduction in the population of our common game animals.

When the parasitism is that of an *animal as parasite and green plant as host,* the parasites are often either worms or insects. Unsegmented worms known as nematodes live in the soil and often parasitize roots of seed plants. They infest the roots of such trees as elm, cherry, and catalpa. When nematodes invade root tissues they interfere with normal root functions and cause dwarfing, foliage discoloration, and wilting. Insects, well adapted for feeding on plants, often extend the herbivorous habit to a parasitic one, living in or on the host plant. Some establish themselves in the tender buds and young foliage, the method used by the spruce budworm. The worm is actually the larva of a small moth that lays its eggs under the bud scales, so the young can feed on the developing foliage. Other insects are bark- and wood-boring beetles that tunnel beneath the bark. The Engelmann spruce beetle lays its eggs in hidden passageways bored in the trunk, where both larvae and adults feed on the inner bark and wood. Numerous sucking insects have become parasites of trees; of these the aphids and scale insects are the most destructive. Aphids induce abnormal growth in the host tissues, causing deformed foliage and spiny or rounded galls such as the cockscomb gall on elm leaves and the spiny gall of witch hazel. Scale insects represent the most highly specialized type of parasitism. The insect has given up its mobility and freedom of movement in exchange for the security of guaranteed meals for its entire life. Some scale insects are encased in a shell-like armor of hardened wax; tightly attached to a tree trunk, they resemble barnacles on a rock. Each scale hides a blind, legless, wingless female whose mouth is permanently attached to the food material of the host's body. Scale insects attack many deciduous and evergreen trees; a notorious species, the San Jose scale, was introduced accidentally into California many years ago and has now become a pest of citrus trees.

In the association of *animal as parasite and animal as host* the parasite is usually an invertebrate, the host a vertebrate. Such parasites are found among the protozoa, worms, and insects. Protozoa enter the host's body in food or water, or through injuries of the skin. The parasitic microorganism feeds on the cells of the host, often destroying them in the process. Trypanosomes are protozoa that live as parasites in antelopes; they also cause sleeping sickness when transmitted to human hosts by the tsetse fly. Other protozoa, known as coccidia, infest many kinds of birds and mammals.

Several kinds of parasitic worms cause diseases of wildlife; among these are the flatworms, roundworms, and segmented worms. The liver fluke is a flatworm that thrives in the bile ducts of ungulates; other flatworms infest the internal organs of many reptiles, amphibians, and birds. The tapeworm is a parasite, highly specialized for existence in the digestive tract of mammals. It has special adaptations for attachment to the lining of the stomach and intestines, and for absorbing food through its body wall. Since this is its only method of obtaining nourishment, the tapeworm cannot live outside the body of the host. Parasitic roundworms include the pinworm of ungulates and the hookworm of many carnivorous species. All parasitic worms consume the reserve food and even the body tissues of their hosts, thus weakening them and blocking the activities essential for their self-maintenance.

Insects are not only masters in the art of parasitizing plants; they display equal ingenuity in getting their sustenance from animal hosts. Few warm-blooded animals escape persecution by the hordes of lice, ticks, fleas, and flies that swarm over their bodies thirsting for blood. These insects use their sucking mouthparts to siphon off the nutritious body fluids of their hosts. Some, like the louse fly of birds and the common body lice of mammals, have hooked appendages to help them get a firm grasp on feathers or fur. Among the common insect parasites are the warble flies of ungulates, the botflies of squirrels, the flesh flies of mink, the screwworm of deer. These insects are doubly dangerous since they also are the carriers of bacterial diseases from one host to another. This is true of the lice that transmit typhus from mammals to man, and of the rat flea that carries bubonic plague from rodents to man.

Adaptation of parasites are directed toward obtaining nourishment without harming the host sufficiently to prevent its continued existence. The longer the host remains alive, the more secure the parasite is in its source of sustenance. When two organisms have been associated in a parasite-host relation for a long time, this goal is often attained. The host adjusts itself to the presence of the parasite so that a mutual balance is achieved, with minimum damage to the host and maximum benefits to the parasite. The introduction of a new parasite into an established wildlife community is another story, however. The host species is usually unable to adjust at once to this new factor in the living environment. All the advantage is on the side of the parasite, which may run wild and completely destroy the host species, but the result is disadvantageous to the parasite also. This happened when the chestnut blight, white pine blister rust, gypsy moth, and Japanese beetle were suddenly introduced into the American environment.

The introduction of domesticated animals may also bring new parasites into a region. This occurred when reindeer were introduced into Alaska and the domestic pigeon into the eastern states. Such immigrant parasites are exposed to no natural enemies in the new environment and so increase rapidly in number, often threatening the extinction of their hosts. This may have been a contributing factor in the disappearance of the passenger pigeon from the American scene, and may present a problem in maintaining a disease-free native caribou population. Such disturbance of the ecological balance by introducing new parasites is a major way in which man has become an upsetting factor in the living environment.

Survival by Predation

A predator is a special type of carnivore, adapted by physique and habits for capturing live prey. Unlike parasites, which are smaller than their hosts, predators are usually larger than their prey. In North America, the predators are chiefly birds and mammals. In both groups the predators reveal adaptations that suit them for an aggressive life.

Mammal predators can be recognized by their large canine teeth, sharp shearing teeth, and strong claws. Since a predator must rely upon keen sight or smell for detection of its prey, these senses are highly developed. The most common mammal predators occur in the weasel, cat, and dog families. In the weasel group, smaller members prey on insects, frogs, fishes, and rodents. Larger species such as the wolverine do not hesitate to attack a deer. Members of the cat family are skilled predators, living on a diet that ranges from mice to deer. In many wildlife communities bobcats, lynx, and mountain lions are the most powerful of the predators. The dog family includes fewer species that are strict carnivores; the wolf, however, is a large predator, challenging the mountain lion for the distinction of being the champion predator.

Bird predators are found chiefly in the hawk and owl families. Their strong beaks are hooked for tearing flesh, their sharply tipped talons are effective in grasping prey. They are powerful and swift fliers, with highly developed vision and hearing. The eyesight of an osprey or a hawk is so keen that these birds can detect a tiny fish or mouse far below them while they soar far aloft.

Few interrelations among animals are so misunderstood as those of predators and their prey. It is understandable that the sudden death of a rabbit when attacked by a bobcat is a shock to a human observer. If one thinks only of the individuals involved, it is natural to consider the killing of a defenseless herbivore by a "bloodthirsty" carnivore a cruel habit. As a result, predators have long been pictured as the criminals of the wildlife community. But ecologists have learned that predation is essential to preserve the necessary balance normal to wildlife populations.

This was clearly demonstrated by events that took place on the Kaibab Plateau of Arizona, north of the rim of the Grand Canyon. Here an evergreen forest occupies more than 700,000 acres, forming a community in which lived such herbivores as mule deer, such predators as bobcats and mountain lions. In 1906 it was made a game preserve, and a policy of eliminating the predators was put into effect in order to increase for hunters the existing herd of 4000 deer. During the following thirty years the toll of slaughtered predators included 816 mountain lions, 863 bobcats, and 30 wolves. The wolves were completely wiped out, the other predators greatly decreased in numbers. This encouraged a rapid growth of the deer herd, which reached 100,000 by 1926. So great a deer population was far in excess of the available supply of plant food. Every bit of grass, shrub, and seedling was consumed. In the two succeeding winters 40 per cent of the deer herd died of starvation.

Ecologists estimated that, at most, the area could support a population of only 30,000 deer. When natural predators were a part of the community, the number of deer remained within the food capacity of the region. When the predators were removed, the deer herd multiplied to a degree harmful even to themselves. A similar sequence has followed predator elimination in other areas. Overpopulation by herbivores, destruction of the plant producers of the community, starvation of the herbivores have been the outcome of such ill-advised policies of killing "savage wolves" and "dangerous mountain lions." Many farmers still shoot all hawks on sight, bounties on bobcats encourage wanton destruction of these useful predators, and airplane "sportsmen" to take a huge annual toll of coyotes. This unwarranted killing is mounting, despite ecological proof of the value of predation in wildlife relations.

Predators play another role, perhaps even more important, in the wildlife community. They prey chiefly on the sick and infirm animals, the weaklings and the cripples. While filling their need to eat, they eliminate animals of inferior genetic stock and as a result tend to improve the race of the animals being preyed upon. This is an important phase of the law of natural selection.

This has been a survey of some of the many interrelations of plants and animals, especially those concerned with the main business of life — obtaining food. Green plants and fungi, birds and insects, predators and prey — together they weave a web of life which enmeshes all the living things that share a common physical environment. The result is a closely knit community. Ecologists have discovered many interesting ways in which this community is organized and how it functions to preserve its unity. We can now proceed to see what these are, to focus on the community rather than the individual organism.

5 / The Biotic Community

OUR FIRST ECOLOGICAL FIELD TRIP to a typical wildlife community introduced us to the role of the environment in determining the kinds of plants and animals which live in an area. Journeying farther afield we broadened our view to survey the different types of plants and animals which live in North America and the many ways in which they become adapted to the physical environment. Then, by focusing our attention on the interrelations of organisms, we discovered how plants and animals become adapted for the main business of life — obtaining food. All of these interactions between wildlife and the physical environment and between different kinds of organisms bring about an interdependent group of plants and animals — the community. We have referred to the wildlife community in general many times in previous chapters. Now, with the knowledge we have of ecological principles, we can take a closer look at the community as an organized unit of nature.

In ordinary language the word community applies to any interdependent group of people, small or large. There is the small community of a fishing village on the coast of Maine, the more complex community of several million people in New York City, and the even more complex American community of fifty states. In the same way, a wildlife community may vary in size from a small woodland pool filled with a few algae, fishes, insects, and snails to a meadow with several hundred species of plants and animals, or to the vast reaches of the prairie with several thousands of species. In ecology, the term community is applied to any group of organisms, irrespective of its size, only if the group possesses a certain organization and is able to perpetuate itself by certain self-sustaining activities. Until now we have been calling such a group of plants and animals a wildlife community. To the ecologist it is a biotic community.

A *biotic community* is an assemblage of species living in a particular habitat and geographic area. But, as has been mentioned before, it is not a chance assortment of organisms. Every biotic community has a complex organization and carries on functions beyond those of the individual members. The organization results in a community structure that binds the members together for their mutual good and in interactions among the members which give each type of community its special characteristics. The community concept is very vital to an understanding of ecology. It brings into perspective the knowledge biologists have acquired about the structure, habits, and distribution of plants and animals. It is not only illuminating to the naturalist — it is of utmost importance to all of us, for man is a member of the biotic community.

Not every group of organisms that live together in the same habitat is a biotic community. A rotting log in the woods may be teeming with microorganisms, beetles, worms, centipedes, and slugs, all living within the restricted habitat of the log and having little to do with the other inhabitants of the woods. But the log will eventually be completely decomposed and the source of energy for all the animals living in the log will vanish. The log population has no organization by which it can perpetuate itself in its special habitat. Nearby may be a century-old maple tree, overgrown with lichens and mosses, serving as a home for many different insects, birds, and mammals. In this group, living within the shelter of the maple as a special habitat, the daily activities often lead the organisms away from the tree. The larvae feeding on the maple leaves develop into butterflies, which get their nourishment from plants in a distant meadow. The birds nesting in the tree, the squirrels retreating to the branches at night, wander far from the tree by day in search of food. And so neither the inhabitants of the rotting log nor the dwellers in the maple tree form a biotic community.

In some cases, a similar area may or may not be a biotic community. This is true of a pool in the woods. If the lives of the inhabitants in the pool can be understood entirely in terms of what happens in the pool, then it would be a biotic community. But if the behavior of the pool inhabitants is conditioned by what happens outside of the pool, then it is not a biotic community. The kind of plants in the pool might be determined by the amount of shade provided by the trees surrounding the pool; many of the insect larvae in the pool may grow into adults that live elsewhere. If in studying the life of the pool we have to refer constantly to events in the woods surrounding the pool, then the pool becomes merely a part of the forest — which is a biotic community.

A biotic community is closely associated with the physical environment of which it is a part. This environment provides the solar energy and the inorganic materials needed to convert sunshine into food. The physical environment and the biotic community it supports form a unit known as an *ecosystem*. This is a self-contained and self-perpetuating system, characterized by a biological equilibrium known in ordinary language as the balance of nature. We shall describe this aspect of the ecosystem in the following chapter. A balanced aquarium is an example of a small ecosystem. In such an ecosystem the water with its dissolved minerals, the light entering the aquarium, the green plants, and the animals can become a biological unit, capable of being sealed off from the rest of the world. A forest and a lake are examples of larger ecosystems, equally self-sufficient as long as their organization remains intact. The largest ecosystem we know is the *biosphere,* which includes that portion of the earth and its atmosphere which sustains life.

Every ecosystem has a basic pattern with four components. These are: (1) a physical environment that provides solar energy and the inorganic materials prerequisite for life; (2) primary energy converters that can transform solar energy and inorganic materials into food; (3) secondary energy converters that transform the energy of plant food into

The physical environment and the biotic community it supports forms a unit known as an *ecosystem*.

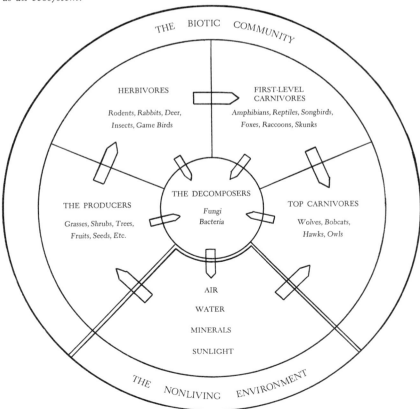

animal tissues and energy; and (4) decomposers that return essential ingredients to the physical environment. These four components make up the blueprint for every ecosystem, small or large, in any region from the arctic to the tropics. The biotic community, which includes the last three of these components, is very sensitive to changes in the physical environment of the ecosystem, which holds a life-and-death control over its very existence. By tampering with this part of the ecosystem, man can cause disturbances that alter or destroy the biotic community, with far-reaching consequences to wildlife and even to himself.

A vivid example of such alteration of the physical environment of an established ecosystem took place in southeastern Tennessee when a copper smelter was erected in the midst of a forest community. The biotic community was dominated by tall tuliptrees, flowering dogwoods, rhododendrons, and azaleas; the vegetation sheltered an abundance of birds and mammals. When the smelter was put in operation, it belched poisonous fumes into the surrounding ecosystem. This new factor in the environment killed all the trees and eventually all the other vegetation. With the primary energy converters eliminated, the rest of the animal life in the ecosystem was doomed. They either died of starvation or went elsewhere to live. First the herbivores abandoned the area, then the carnivores, as their prey disappeared. With the loss of the vegetation, the unprotected soil was washed away by the rains, and little water soaked into the ground. Brooks and springs dried up. Rolling hills became barren ridges and gullies as humus-rich soil was carried away to be deposited as silt in distant reservoirs. A lifeless desert took the place of a once populous biotic community.

The Organization of the Biotic Community

Given a stable physical setting, every biotic community follows the same general organization, and in each certain jobs must be done for the good of the community as a whole. Each job — or work assignment — is known as a *niche*; each niche carries out the function of a particular component of the community pattern. Food production by green plants is the niche where primary energy conversion is accomplished. Secondary energy conversion usually involves two niches: that of plant-food consumption by herbivores and that of animal-food consumption by carnivores. The decomposers fill the niche in which the carbon and nitrogen cycles are maintained by bacteria. These essential niches are repeated in every community; their functions are the same in each, but the organisms that fill the niches are not necessarily the same. The

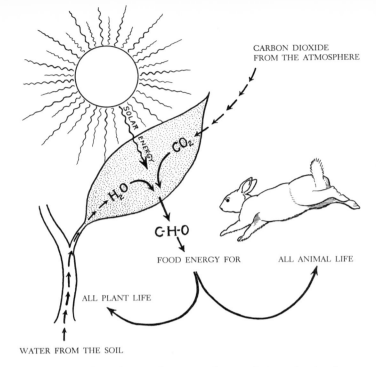

Photosynthesis — food production by green plants — brings about primary energy conversion in the ecosystem.

niche should not be confused with the habitat of an organism. The special type of physical environment in which a plant or animal lives is its habitat; the special kind of work done by the plant or animal, its role in the community activities, is its niche. The habitat can be likened to an organism's address in the community, the niche its profession.

The niche of the food producers is a very essential one. Without it, all other life in the biotic community comes to an end. In a prairie community this niche is occupied by grasses; in a tundra community by lichens; in a forest community by trees. All the occupants of this niche are not of equal importance in community life. Some are dominant species by virtue of their greater food production, of their more effective shelter for smaller organisms, or their greater suitability for living under the existing community conditions. The dominant plants determine the characteristic of the community: conifers in a Canadian forest, oaks in an Ohio forest, mesquite in a Texas grassland. Often the dominant species moderate the physical nature of the environment so that other less vigorous or less well adapted species can become members of the community.

The niche occupied by the herbivores makes possible all other animal life of the biotic community. Herbivores are the middlemen who deal directly with the wholesalers (food producers) and in turn retail the converted food energy to the carnivore consumers. The type of herbivore which fills this niche depends upon the kind of vegetation in the food-producer niche: prairie dogs and antelopes on grasslands, moose and varying hare in conifer forests.

Carnivores fill two niches in the biotic community. A group of first-level carnivores obtain their food by eating herbivores. This niche is occupied by different species, depending upon the types of plants and animals filling the other niches. In the prairie dog–grassland community the carnivore may be a coyote. In the spruce-hare community it may be a bobcat. A second-level carnivore group, often referred to as "top carnivores," fill the niche of the large predators that feed on smaller carnivores. They may be wolves or mountain lions, depending upon the species making up the other niches and the geographic location. At each succeeding niche the food energy passes from one type of organism to another; it originates in the food producer and terminates in the top carnivore.

Some niches are filled by special types of consumers, such as parasites and scavengers. In the decomposer niche there may be a variety of fungi as well as bacteria. Whoever the occupant of a niche may be, his permanence is by no means guaranteed. Often the original occupant is displaced by a new and invading species. If the new species is better adapted to do the job, it eventually takes over and dispossesses the former occupant. This has often occurred when an aggressive species, usually introduced by man, moves from one community to another. Outstanding examples are the invasion of native songbird niches by the English sparrow, or of native grasslands by the dandelion.

The Classification of Biotic Communities

Climate determines the largest of the ecological units used in determining biotic communities. These major units are known as *biomes*. Each is a community of large size, characterized by the distinctive life form of the plants that dominate the community. This life form is an adaptation shared by a number of species living under the same climatic conditions. It includes the length of life of the plant (whether annual or perennial), nature of the stem (whether herbaceous or woody), the arrangement and character of the leaves (deciduous or evergreen, broad-leaved or needle-leaved). The life form of the dominant plant members of the community gives the name to the biome. Thus we find the deciduous forest biome of the eastern states, and the grassland biome of the central states. The life form of an animal is also the result of all its adaptations to the physical environment of the ecosystem, but it is also determined by the vegetation that provides food and shelter. Many mammals are arboreal, a life form suited to climbing in the trees, to be seen in the raccoons and squirrels. Where the biome lacks trees, arboreal animals are consequently absent. There are six major biomes in North America, of which more will be said in later chapters.

Each biome can usually be subdivided into smaller biotic communities determined by soil and surface features, by the amount of moisture in the ground, and other factors of the physical environment which may vary within the climatic boundaries of the biome. These are known as *associations*. Each association is made up of certain species that dominate the community. If we explore the deciduous forest biome from West Virginia to Minnesota, we find that broad-leaved angiosperm trees occur throughout the area; but in West Virginia it consists of a tuliptree-oak association, whereas in Minnesota it is a maple-basswood association. In many cases the animal life of each association is also distinctive.

Smaller self-sustaining communities are found within the association. On our ecological field trip through an eastern deciduous forest we encountered some of these smaller ecological units. In the low wet ground we found a swampy-pond community of cattails, alders, red-winged blackbirds, and amphibians. On high and dry rocky ridges another local community occurred, dominated by pines and oaks, with such mammals as live among trees of this type. These small communities are sometimes called societies. One society may thrive on the southern slope of a hill, another on the cooler, moister northern slope. They will, however, consist of plants and animals with the life form suited to the association and the biome of which they are a part.

Such is the framework within which the community functions. We can now continue with the activities that take place within this organization, and which present a dynamic picture of the biotic community in action.

6 / The Biotic Community in Action

ONE OF THE UNIVERSAL CHARACTERISTICS of life is its organization. This organization reveals itself at different levels of complexity and can be pictured as a "biological spectrum." At each level certain activities become possible because of the degree of organization present. At one end of the "spectrum" is *protoplasm*, the organized molecular basis of life. Protoplasm in turn is part of another organization level, the *cell*; in higher forms of life, cells are organized into a *tissue*. The biosciences of cytology, cellular physiology, and histology study life at these organization levels. At still higher levels we find the *organ*, made up of tissues, and the *organism*, an integrated and interdependent group of organs. Traditional biology (morphology, anatomy, physiology) studies life at the organism level.

We are accustomed to think of the organism — whether a plant, an animal, or man — as the highest level of organization. The ecologist, however, goes on to consider even higher levels of complexity. He considers that all the individual organisms of a species form a *population*, with distinct characteristics and functions apart from those of the organisms of which it is composed. A group of populations living in the same habitat or area forms a still higher level of organization, the *biotic community*. We know that an individual organ, such as an arm or leg, cannot long survive without its organism. Similarly, the organism cannot live long without its population, or the population without its community. At the opposite end of the "spectrum" from protoplasm is the ecosystem, a complex and highly organized group of biotic communities. Here too the organization brings with it a mutual dependence. The community cannot long exist without the constant flow of energy in the ecosystem.

The activities that take place at the community and ecosystem levels involve two basic concepts: that of a balanced energy flow, which makes the ecosystem a self-contained unit, and that of homeostasis, which creates stability in the community. *Homeostasis* (from two Greek words meaning "similar" and "standing") is a well-known concept of biology

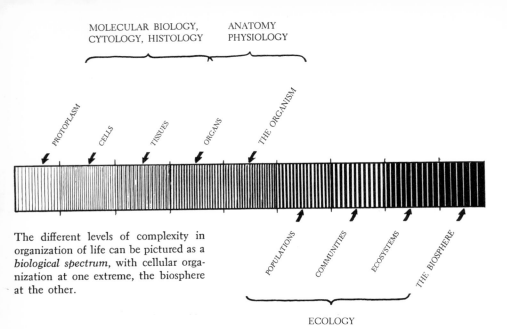

The different levels of complexity in organization of life can be pictured as a *biological spectrum*, with cellular organization at one extreme, the biosphere at the other.

at the organism level. It means the ability of the body to preserve its normal condition in the face of constant change, to maintain an internal biological equilibrium. Homeostatic forces enable a warm-blooded animal to keep a constant body temperature, or a diseased animal to bring its body defenses into action against invasion by microorganisms. Like the organism, the community is constantly exposed to forces tending to disrupt its organization. But it too has built-in checks and balances that act as homeostatic forces, creating stability of the community in spite of changes that may take place in the ecosystem.

The Energy Flow through an Ecosystem

In describing how an organism works, it is often helpful to compare the living mechanism with a mechanical device. This is especially true when considering the energy aspects of life, since transformation and utilization of energy are carried out in the nonliving world by machines. Thus it is possible to compare the heart and circulation to a pumping system, and nervous coordination to a computer. A similar approach can be used in describing the flow of energy into and through the various niches of a biotic community, by comparing it with the input and outgo of energy through a mechanical system.

Energy Flow through an Ecosystem. A surprisingly small amount of the total solar energy entering an ecosystem is utilized by the biotic community. It is estimated that the plant producers convert only 1 per cent of the total solar energy entering an ecosystem. The remainder passes out of the ecosystem as heat loss. Yet this small per-

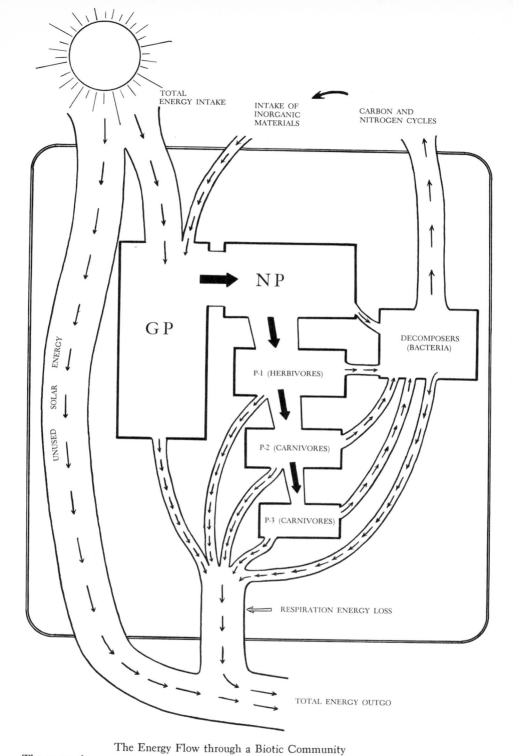

The Energy Flow through a Biotic Community

The rectangle represents the boundaries of the biotic community, into which energy enters from the available solar radiation. A portion of this energy input, via green plants, becomes food (NP) which passes the energy on through successive levels of consumers (P-1, P-2, P-3). Outgo of energy from the community is largely the result of plant and animal respiration.

centage is sufficient to power all the animal activities of the entire community, which may number thousands of species of insects, worms, frogs, snakes, birds, and mammals — including man. Tracing the flow of this energy through the ecosystem we find that only part of the energy transformed into plant food becomes available for other niches in the community. Some of the primary conversion of solar energy is used in plant respiration and the living activities of the plants themselves. Thus the *net* primary food production, indicated by NP in the diagram, is less than the *gross* primary food production, indicated by GP in the diagram. The difference between the two represents the energy lost as the food passes from one energy level to the other.

As the food passes through successive consumer niches, indicated by P-1, P-2, and P-3, smaller and smaller amounts of the original net primary energy reserve becomes available. This flow of energy through the ecosystem therefore obeys the second law of thermodynamics, which applies to living energy systems as well as mechanical ones. In simple language, according to this law, energy passes from a concentrated to a dispersed state. At each transfer of such energy, as from herbivores to lower-level carnivores to top carnivores, some is lost as heat through respiration.

It is obvious that a relative balance in numbers of individuals must be maintained between the organisms that supply the energy and the population in the next "higher" niche whose continued existence depends upon receiving energy in an uninterrupted flow. The stability of this energy-transfer process is secured by "safety valves" within the biotic community. One such self-regulatory device is the automatic increase in mortality within a species when its population exceeds the limit set by the ecosystem. In other words, overpopulation at any level threatens the energy balance inherent in the available food supply of the community, and upsets the energy flow. When this happens, the self-regulatory mechanism makes use of other forces in the ecosystem, those like parasitism and predation, to restore the balance. This keeps the population of a niche within the natural limits. The incident of the predator-herbivore situation in the Kaibab Plateau community illustrated this principle.

Food Chains. The flow of food energy within a biotic community, as it passes from one niche to another, creates a food chain. Each food chain is the transfer of net primary energy through successive levels of consumers. Some of these energy transfers consist of two links, others of three, and some have as many as five links. The shorter the food chain, the greater the amount of primary energy available to the last member of the chain. The herbivore in a two-link chain is in the best situation. Less fortunate in securing energy is the top carnivore, which is the last in a longer food chain. A *parasite food chain* usually consists of only two links, and leads from large animals and plants as hosts, to smaller organisms, usually invertebrates and nongreen spore plants. A *saprophyte food chain* is also ordinarily a two-link chain, leading from nonliving organic matter to microorganisms such as bacteria, and to fungi such as mushrooms.

The predator food chain is a type of energy transfer of common occurrence in nature. The predator food chain is a short one in regions where living conditions are severe, so that only a few species occupy each niche. In the Arctic, for example, a three-link chain leads from lichens → caribou → wolves. Another three-link chain is found on the prairies: grasses → bison → wolves (or man). In temperate forests the predator chain may consist of four links. Such is the spruce tree → wood-boring beetle → woodpecker → hawk food chain, or the oak → chipmunk → weasel → bobcat chain. Where living conditions create an abundance of life, as in the tropics, the number of interacting predator, parasite, and saprophyte relationships result in complicated and numerous food chains. Food chains become complicated also by the omnivorous habits of many carnivores which have learned that survival in the struggle to get energy is easier if the diet is not limited to too few sources. The direction of the predator food chain is usually determined by the size of the animal. There is a minimum size below which the prey cannot furnish adequate food energy to the carnivore; there is also a maximum size above which the prey cannot be overpowered.

Life Pyramids. The concepts of energy transfer and food chains lead to another aspect of community organization. This involves the numbers of individuals occupying each niche. If the numbers of individuals of a species, or their total weight, are represented for each niche, a pyramid results. At the base are the green plants that provide the primary energy for the entire community. Since the plants must use some of this energy, the production must be greater than the total energy consumption at the second transfer level. Such consumption can be measured either in the total number of individuals or their total living weight. In the same way, energy production at the second level must exceed that needed at the third. Food consumption is proportionate to the size of the population; and so the total numbers of carnivores (or their living weight, known as "biomass") must be less than that of the herbivores supporting them. At the top of the pyramid are the few large predators that can be supported by all the niches below them. If the amount of energy is reduced at any level, it affects the lives of all the populations above that level. The nearer to the base a reduction in food transfer takes place, the more disastrous it is to the entire biotic community.

A simple pyramid can be worked out by patient count of all the organisms in a definite area of the community. It involves careful identification of all the individuals in each niche. This has been done for an acre of grassland, with the following results. At the base of the pyramid were 5,842,424 grass plants. Their net energy passed on to 708,624 small invertebrate herbivores, chiefly insects. These herbivores in turn supported 354,904 carnivores, mainly spiders, ants, and beetles. At the summit of the pyramid were three top carnivores. One can readily imagine what would happen to the two moles and the bird, which were the top carnivores, if a sudden population growth increased the number of herbivorous insects — for instance, the locusts — so that all the grass was destroyed.

STABILITY VERSUS CHANGE IN THE COMMUNITY

A biotic community may seem a permanent feature of the landscape, but it is the scene of many changes. This is one of the fascinating aspects of nature, and one of the most dynamic phases of ecology. Some of these changes within the biotic community take place at short time intervals and occur in daily or seasonal cycles. Such are the alternation of day and night, and the changes accompanying summer and

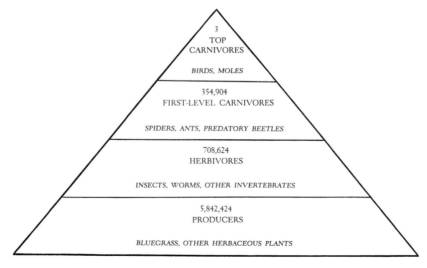

The concepts of energy-transfer and food chains lead to another concept, that of a *pyramid of life*, with producers forming the supporting base and second-level carnivores at the apex.

winter. Another type of change, known as succession, requires much longer periods of time, often hundreds or even thousands of years. Since the change covers such a span of time, one could observe it only through the use of time-lapse photography, with exposures made every ten or twenty years. In such a successional change, one type of biotic community is gradually replaced by another. This happens when a burned-over conifer forest becomes aspen woodland, or when a lichen-covered rock outcrop becomes a spruce forest.

Cyclic Changes in the Biotic Community. The changes occurring in a biotic community have their counterpart in a human community. Cyclic changes in a town can be seen in the contrast between daytime and nighttime activities. During the day businessmen and housewives hurry through the streets, streets are thronged with traffic, stores and factories are filled with customers and workers. During the night factories and stores become silent, the streets deserted. Nightwatchmen replace workers; milkmen make their rounds and policemen patrol their beats.

In the biotic community the change from day to night also affects the activities carried on in the various niches. With the coming of darkness the green-plant producers cease work. In the animal niches a night shift takes over from the day shift. Now it is the turn of the night-flying moths to assume the task of flower pollination. Nocturnal herbivores and carnivores replace those which work by day. Sunlit fields resounded with crickets and frogs, but at night the chorus is resumed by toads and katydids. Thrushes and hawks, busy by day, yield the night to whip-poor-wills and owls. Squirrels and chipmunks disappear, and skunks and raccoons become active. Since some animals are diurnal and some nocturnal, their periods of feeding do not coincide. Because of this alternation of food-transfer activities, the biotic community can support a greater number of species in each niche, with less competition among those having similar methods of obtaining their food.

Seasonal changes also take place in both the biotic and the human communities. Some members from each migrate to the Southland as winter approaches, and in summer return to the cooler northern regions. Agricultural workers may move from one community to another as crops need harvesting. Outdoor activities lessen in regions with cold winters, and indoor ones take their place. The alternation of seasons in temperate regions likewise brings periodic changes in the biotic community. In a deciduous forest the trees become leafless, the absence of insects creates a great silence, amphibians and reptiles have withdrawn to their winter retreats, and many birds have gone south for the winter. Of the mammals that remain, few are about; many, out of sight, are sleeping or hibernating. Community energy production and transfer is at its lowest ebb in the cold season.

With the coming of spring, the energy producers return to action. Leaves reappear on the trees, new plants spring from the earth, grass becomes green. This spurs increased activity in the upper levels of the life pyramid, as food chains are again established; the community hums with activities that reach their climax in summer. The abundance of food spurs the reproduction of plants and animals, adding to the community population. When autumn arrives, energy production at the base of the pyramid begins to slow down. Autumn colors foretell the death of the food-manufacturing organs and the leafless winter season returns. During this seasonal change some niches are occupied by short-term residents that live but a year, taking part in the community activities only once in each cycle. These are the annual plants and the majority of insects. Other residents have a longer lease in their niche and reappear each spring: the perennial grasses, shrubs, and trees, as well as the vertebrates. A similar cyclic change takes place in desert communities, where alternation of a wet and dry season brings about a corresponding change of pace in community life.

Successional changes in a biotic community also have their counterpart in a human one. Small pioneer settlements grow into big cities; other communities, once populous, become ghost towns. The native population of a community may be crowded out by an immigrant population with willingness and ability to take over jobs previously performed by the original inhabitants. New families move into town, old families die out as change of environment takes place. We shall see that comparable changes take place in the history of a wildlife community.

At first all the land surface of the earth was a rocky crust, barren of life. How did the first biotic communities become established, and what were they like? Can succession be seen as it takes place today? What happens when a community is destroyed, and how is it replaced by another community? These are a few of the questions on which the concept of succession throws some light.

Succession consists of a series of stages, each stage having populations filling the basic niches, which are adapted to live under the prevailing conditions of the ecosystem. In the colonization of a barren area, the first community is made up of pioneer species capable of filling the available niches. The food producers appear first; they are usually simple spore plants and herbaceous perennials soon becoming abundant enough to support the pioneer herbivores, frequently small rodents. The pioneers alter the living conditions so that often they must give way to another community, one better suited to the new environment. This may include shrubs and small woody plants, birds and additional mammals. This community in turn is replaced by another, each changing the environment and, in so doing, paving the way for its successor. In the process it often seals its own doom. Finally succession reaches a type of community able to maintain itself indefinitely because it is in equilibrium with the climate of the region. This is known as a *climax community*; it is the fullest expression of life in a particular ecosystem. Homeostatic forces will enable it to survive as long as the climate and topography remain the same, and if it is not disturbed by man.

The replacement of one community by another, with an attendant succession of species in the various niches, is brought about by three different natural causes: changes in climate, changes in physiography, and changes initiated by the biotic community itself. Climatic succession is the most far-reaching in its effect on the ecosystem; it also spans the greatest length of time. Physiographic succession takes place if the climate remains constant but the earth's surface changes in a way to alter the physical environment of the ecosystem. This too takes time, though less than climatic succession. Ultimately there is biotic succession, resulting from the activities of the plants and animals in changing the ecosystem. This last type of succession is the result of forces inherent in the community itself; it is often noticeable within the lifetime of a human observer.

Climatic Succession. The climate of any region is far from constant throughout long periods of time. It changes as a result of major alterations in the atmosphere and in the earth's crust. In North America the advance and retreat of glaciers over the northern part of the continent have had profound effects on the climate. Because of the varying climate, the biotic communities as they now exist were not necessarily present a century or ten centuries ago. Searching for proof that climatic succession has taken place, biologists have accumulated evidence from various sources, particularly in traces of plants and animals formerly inhabiting the ecosystem. In Indiana bogs, left by the retreat of the last glacier, they discovered sediments at the bottom of the bogs containing successive pollen deposits from the trees that grew around the bog. The lowermost pollen deposits included pollen from conifers; the uppermost layers, from deciduous trees such as oaks and hickories.

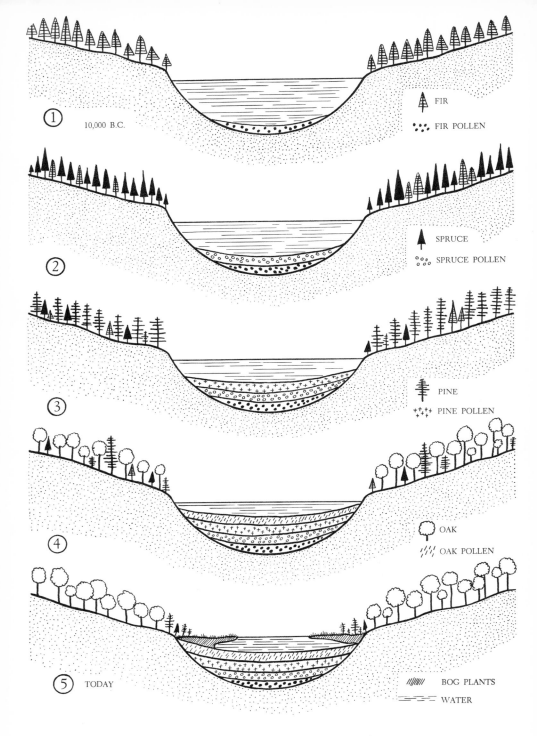

Climatic succession is revealed by studies of pollen deposits in bogs. (1) After retreat of the glacier in Indiana, balsam fir appeared. (2) Then the fir moved north and white spruce came in. (3) When the spruce moved north, as the climate warmed, pine became established. (4) Finally, the milder climate favored the growth of oaks, and most of the pines moved north. (5) The oak forest has remained to this day. Some pine and spruce lingers on the bog mat, which is gradually covering the entire lake. Adapted from M. T. Watts, *Reading the Landscape*.

THE BIOTIC COMMUNITY IN ACTION / 87

From these the story of a climatic succession which extended over a period of 10,000 years could be reconstructed. The first pollen of the bottom sediments, that of balsam fir, indicated a community living in a cool moist climate. The next pollen layers contained more spruce than fir pollen, suggesting a climate less moist but still cold. Successive pollen layers revealed a community of pines, typical of a drier climate. Finally, an abundance of oak pollen was evidence of a warm dry climate that encouraged the development of a community much like the one

Physiographic succession can be seen in the stages of valley formation by an eroding stream in an eastern maple-beech forest. Adapted from M. T. Watts, *Reading the Landscape*.

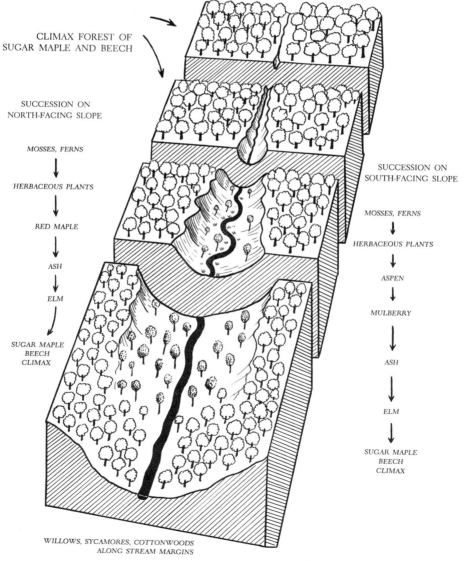

present in the area today. Fossil skeletons in other bogs disclose the animal members of these ancient communities. Bogs in New York State have yielded remains of mastodons and mammoths, prehistoric mammals of the early stages in climatic succession.

When the earth's surface is altered, *physiographic succession* occurs. A stream erodes a tiny gully through an established climax community; it then becomes a deep ravine, and in time a broad alluvial valley. This brings about a succession of biotic communities reflecting the changed conditions of the environment. If this erosion happens in a beech-maple forest, the sequence of succession begins in the ravine with pioneers like mosses and ferns clinging to the shaded and damp sides of the ravine. As the valley widens, light-tolerant shrubs and trees such as hemlock appear. In the wide alluvial valley the hemlocks and mosses give way to an open woodland of ash, poplar, and willow. Alders line the stream, young sugar maples and beeches advance down the slopes. When the broad valley finally provides an environment similar to that of the original community, the beech-maple climax becomes re-established. Each of these stages in the physiographic succession has its distinctive animal members.

Biotic succession is more readily observable. In an ecosystem which is in the process of becoming a climax community, the physical environment may be bare rock, or windblown sand dunes, or small bodies of water. In all of these the pioneer stages will be different, but all lead to the same climax if there is a uniformity of the climate over the area. By carefully exploring the ecosystem, one can find evidence of different stages in succession at different locations. From these the sequence in succession from its earliest beginnings to its climax can be inferred.

Succession beginning with bare rock can be seen on any of the many rocky ledges found throughout northern New England. The first signs of a biotic community are advancing growths of crust lichens encroaching on the barren spaces. As the rock surface weathers and disintegrates under lichen activity, an accumulation of inorganic and organic debris provides a shallow soil for erect lichens, mosses, and xerophytic grasses. The vegetation mat formed by these low-growing plants soon covers the original-crust lichen community; in the substratum it provides, seeds of perennials can germinate. As more soil accumulates, small shrubs appear. These encircling communities grow toward the center of the original area of bare rock, which soon disappears beneath the plant cover. The soil is now deeper and retains more moisture; seedlings of spruce and pine form a beginning forest growth. In the final stages,

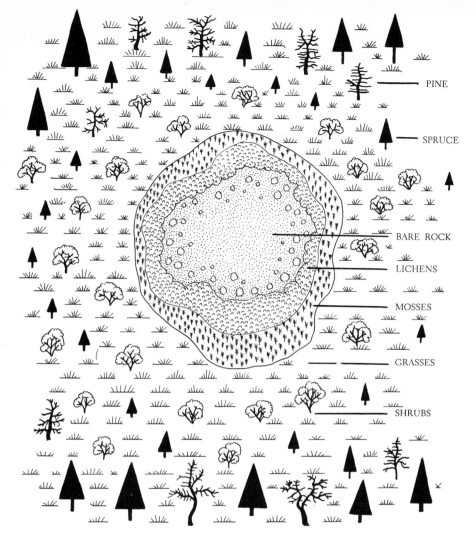

Biotic succession on bare rock in northern New England begins with lichen-moss pioneers and ends with a pine-spruce climax forest.

these trees with other less dominant species like birch and alder form the climax community. Here and there may be vestiges of the pioneer stages — lichen beds, mossy glades, grassy and shrubby openings in the forest. Paralleling these changes in the vegetation are changes in the animal occupants of the herbivore and carnivore niches.

Another type of biotic succession, originating also in xerophytic communities, can be seen in the colonization of sand dunes. Here the pioneers are dune grasses and hardy shrubs whose root systems can anchor the vegetation in the shifting sand, and whose tough leaves reduce transpiration caused by the constant winds. These pioneers

Biotic succession can also convert a hydrophytic community, such as a pond, into a mesophytic forest. (1) Submerged vegetation; pond beginning to fill with sediment. (2) Emerging vegetation; pond becoming shallower. (3) Swamp stage; shrubs and trees moving into the disappearing pond. (4) Beech-birch-maple forest, almost entirely obliterating vestiges of the original pond and its inhabitants.

stabilize the soil, encouraging invasion by sand cherries and dune willows. Finally the larger trees, cottonwoods and lindens, appear and the dune becomes an open woodland. In time the substratum develops a moister soil, so that a mesophytic forest terminates the succession. It is made up of maples and oaks, together with some pines and junipers.

A third type of biotic succession begins with hydrophytic vegetation of ponds and swamps. Most ponds are temporary features of the landscape, destined to be replaced eventually by land and a climax land community. In a typical pond situated in an eastern forest, the succession usually takes a course leading from completely aquatic plants and animals to terrestrial ones. The pioneers are the submerged pondweeds rooted on the bottom. As they die, their remains accumulate and build up sediments that make the water more shallow. Now the community includes floating aquatic plants, rooted in the bottom but with long stems which, as in pond lilies, can support leaves on the surface of the water. Slowly the pond fills up with bottom deposits until the water is only a few inches deep. A new community appears which is a swamp stage with cattails, rushes, and sedges. As they in turn add to the soil, the swampy ground becomes drier and a community of woody shrubs — willows and alders — becomes established. At last, far back from what remains of the open water in the pond, a mesophytic forest develops, dominated by maples, birches, ashes, and beech. In time, land will eliminate the pond completely and the climax community will take its place. Animals, too, have taken their place throughout this third type of biotic change: muskrats and loons, then red-winged blackbirds and frogs, finally thrushes and squirrels.

These examples of biotic succession are the result of the activities of plants and animals, which bring about changes in the ecosystem: filling in ponds with organic debris, adding nutrients and a stable substratum to shifting sand, transforming bare rock into soil and a vegetation cover. Biotic succession is sometimes, however, the result of interference by man. When a climax forest has been cut down and the land cleared for agriculture, the farmer fights a constant battle against grasses, weeds, and seedlings that strive to re-create the original forest by succession. When such a field or pasture is abandoned, succession begins at once, stage by stage. This process can be seen where farms have outlived their usefulness and nature reverts to its original state. Fire, excessive use of chemical sprays, fumes, overgrazing — all bring about a change in the climax community or a modification of whatever stage in the succession is taking place. But given opportunity, biotic succession through its many stages will bring about the return of the community best suited to the total environment, a duplication of the former climax community. This, moreover, will take many years.

Near New Brunswick, New Jersey, is a small wilderness tract known as Mettler's Woods. Although in the midst of a highly populated part of the country, it has been spared destruction by being owned continuously since 1701 by one family that has refused to allow the land to be lumbered, converted into agricultural land, or otherwise disturbed. It is a deciduous forest community much like the one we have explored. A study made by Rutgers University biologists estimates that were Mettler's Woods destroyed it would eventually be replaced, but at least two hundred years would pass before the area would return to its original state. For the first few years, after the woods were lumbered over, the area would be a stumpy field overgrown with ragweed, goldenrod, and grasses. During the next ten years a few woody shrubs would appear: bayberry, cherry, and viburnum. Among them would be a few red cedars, trees that can grow in dry and often sterile soil. For fifty or sixty years the cedars would thrive and multiply, forming an open woodland. During these years the soil would become richer and deeper, providing a suitable home for a few seedling elms and red maples. After another fifty years the stage would be set for occupation by seedlings of the same species of trees as dominated the mature forest. But these oaks and hickories grow slowly, so it might be a hundred years before they formed a dense forest canopy. Within the next hundred years, beneath these deciduous trees would develop the rich variety of the smaller shrubs, flowering plants, ferns, and mosses. With food and shelter again available, herbivores would establish themselves in the community, and soon the carnivores would appear. Thus three centuries might well elapse before Mettler's Woods would again be a wilderness community — and there is little that man can do to speed up this recreation of a unit of nature.

II
The Biomes of North America

7 / Life and Its Physical Environment in North America

In exploring a small portion of the earth's surface such as the wilderness area described in Chapter 1, we have seen how the ecological view gives meaning to what otherwise might seem a haphazard association of plants and animals. On trips farther afield, driving the length and breadth of our country, we can gain an even better understanding of the American landscape and its wildlife by observing it through the eyes of an ecologist.

One trip can take us along the Atlantic Coast on Route 1, as it leads from the Canadian border to the Gulf of Mexico. For much of this distance the route follows the Atlantic shoreline of North America, from the rugged coast of Maine to the coral keys of Florida. At the northern terminus of the route snowdrifts still linger along the roadside in April. The scattered deciduous trees are bare, but great numbers of spruce and pine form a living green landscape. Few animals are in evidence. We may see a red squirrel scampering beneath the trees, or hear a white-throated sparrow cheering us on our southward journey. In southern New England the conifer forests become thinner and deciduous maples and oaks dominate the roadsides as we bid farewell to the last of the winter's snow. Robins are flying northward and a woodchuck basks in the spring sunlight.

After leaving Washington, D.C., we notice that the grass is greener, the red maples are in full bloom. Rolling, fertile hills take the place of rocky hillsides. Deciduous woods of maple and oak have an undergrowth of blossoming redbud and dogwood. The woods are alive with mockingbirds and cardinals; opossum and fence lizards replace the red squirrels and woodchucks of New England. In the Carolinas the air becomes moister and milder, the woods more populous with insects, songbirds, and mammals. Groves of glistening longleaf pines mingle with the red maples and oaks. Impenetrable jungles of palmetto and bald cypress grow in the low swamplands; here the trees are draped with Spanish moss and the black swamp waters swarm with frogs and turtles.

In Georgia and Florida, live oaks with evergreen foliage are added to the scene, as are the glossy-leaved magnolias and a tangle of flowering vines. Cabbage palms give a tropical touch to the roadsides. Agile lizards and bright green anoles seem to be everywhere, on fences, roadside banks, and tree trunks. Long stretches of coastal grassland alternate with "islands" of oak and pine. At Key West, the southern terminus of Route 1, we find ourselves in the tropics. Mangroves and papayas, royal and coconut palms border the highway. Deep in the swamps live alligators and crocodiles; white egrets and ibis roost in the trees bordering the roadside canals. The palm groves of the Southland are a far cry from the spruce forests of Maine.

A different view of the continent is revealed by a journey along Route 40, which spans the distance from Baltimore to San Francisco. It is early summer when we leave the eastern lowlands and the deciduous forests, freshly green with the new foliage of maple, elm, and oak. The highway leads westward through the mountains of Maryland, winding through woods filled with songbirds, raccoons, and squirrels but with fewer amphibians and reptiles than we encountered along Route 1. At the crest of the Appalachian Mountains, scattered groves of spruce remind us of the sea-level forests of Maine. West of the mountains the forests grow less dense; as the highway crosses Ohio into the flat plains of Illinois, the trees become smaller and the forests become open groves of scattered oaks and hickories. Approaching St. Louis, on the banks of the Mississippi River, we find the countryside a mixture of grassland and woodland.

West of the river the highway imperceptibly climbs a gently sloping plain across Kansas. Forests have now completely disappeared and have been replaced by prairie grassland, where there are coyotes and prairie dogs instead of squirrels and foxes. In Colorado the road rises more steeply, the grasses become sparser, and prickly pear cacti add a novel touch to the landscape. At Denver the highway has climbed to a mile above sea level; ahead is the soaring backdrop of the Rocky Mountains, where again we meet the forests. Conifers clothe the slopes with a rich dark green, but the trees are Engelmann spruce instead of red spruce, ponderosa pine instead of the eastern white pine. New birds and mammals greet us in bewildering numbers.

At Salt Lake City the highway leads away from the Rocky Mountains. Behind are the deciduous forests and the prairie, the evergreen-forested slopes of the mountains. Ahead in Nevada is a plateau of alkali flats and arid slopes dotted with sagebrush and stunted juniper. At Reno another mountain barrier confronts us: the snow-capped Sierra Nevada. The highway climbs up to a mountain pass where the air is cool and refreshing as we pass through forests of ponderosa pine. The shimmering desert lies far below and summits capped with snow tower above the timberline conifers. From this point the route descends to another parched lowland: the interior valley of California. Earlier the grassy hillsides were green from the spring rains, but now they are a golden brown. The scattered trees are gray-green digger pines and live oaks, and the open parklike landscape is dotted with spherical cushions of the bayonet-leaved yuccas. Again the highway winds up through low wooded hills, and then makes a final plunge to the shores of the Pacific Ocean at San Francisco. Route 40 has revealed another cross section of our continent.

The Surface Features of North America

The journeys over Routes 1 and 40 have given us a general idea of the topography of North America. Now let us imagine we are many miles out in space, high enough to see the entire continent spread out beneath us. Below we see a wedge-shaped piece of land, broader along its northern limits than at its southern boundary. The eastern and western margins are elevated to form mountainous rims, one bordering the Atlantic Ocean, the other the Pacific. Between them is a broad central lowland that stretches from the Arctic Ocean to the Gulf of Mexico.

The eastern mountainous rim is known as the *Appalachian Highlands*. This begins in New England and extends southward to Tennessee and South Carolina. The western rim of the continent consists of two parallel ranges: the Pacific Coast Ranges and the Rocky Mountains. Together they form a broad and rugged backbone to North America, extending its entire length from Alaska to the Rio Grande River. This western highland region differs from the eastern in several important respects. It is a much broader upland region, a thousand miles or more from east to west compared to a few hundred miles in the Appalachian Highlands. The mountains also attain much greater heights, with many peaks above 10,000 feet in elevation. In addition, the mountainous rim plunges directly into the Pacific Ocean with no intervening coastal plain.

The *Pacific Coast Ranges* begin in the Far North in the Alaskan Range where Mount McKinley, the highest peak on the continent, reaches a snow-capped 20,300 feet. Southward the Pacific Coast Ranges

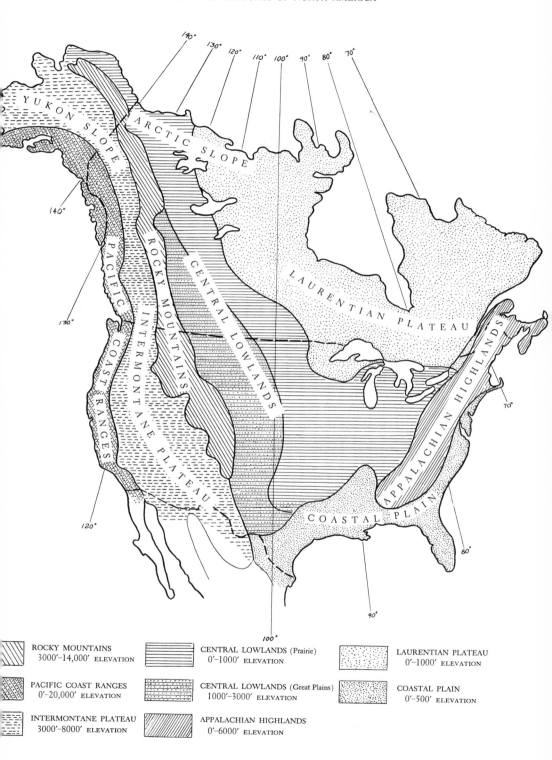

diverge to form two more or less parallel mountain systems. On the seaward side are the low coastal mountains of western Washington, Oregon, and California. To the east of these lie the Cascade Ranges, extending to northern California.

The eastern portion of the mountain system consists of the *Rocky Mountains*, which originate in Alaska and extend through western Canada into the United States, to Colorado. Between the Pacific Coast Ranges and the Rocky Mountains is the *Intermontane Plateau*. A northern portion, or Columbia Plateau, is found in eastern Washington and Oregon; a southern portion, or Colorado Plateau, extends over Utah, Colorado, New Mexico, and Arizona.

The remaining portion of North America is a central broad lowland that occupies about one third of the continent. Its northern portion forms the *Laurentian Plateau*, which slopes gradually to the Arctic Ocean; elevations range from sea level to a few thousand feet. South of the Laurentian Plateau the *Central Lowlands* forms a north-south trough between the Appalachian Highlands and the Rocky Mountains. Another smaller lowland, the *Atlantic Coastal Plain*, separates the Appalachian Highlands from the Atlantic Ocean; it continues southward from New Jersey in a wide arc to the Gulf of Mexico. The entire peninsula of Florida is coastal plain recently emerged above sea level.

This, in brief, is the topographic diversity that forms the background for American wildlife, presenting many types of habitat from coastal swamps to barren mountain summits, from rocky canyons to fertile valleys, from arid plateaus to lush lowlands. It has given North America its unique scenery and at the same time numerous regions whose distinctive surface features each present an environment with different factors capable of affecting life.

THE CLIMATE OF NORTH AMERICA: TEMPERATURE

The climate of our continent is as varied as its topography. One reason for this is the great range of temperatures existing in various parts of the country. Temperatures have been recorded of a low of $-76°$ F. at Yukon Valley, Alaska, to a high of $134°$ F. in Inyo County, California — a range of more than 200 degrees. Few states escape subzero temperatures in winter, and surprisingly few states escape occasional summer temperatures of $90°$ F. or above. Such temperature extremes require special adjustment for plants and animals living in these regions.

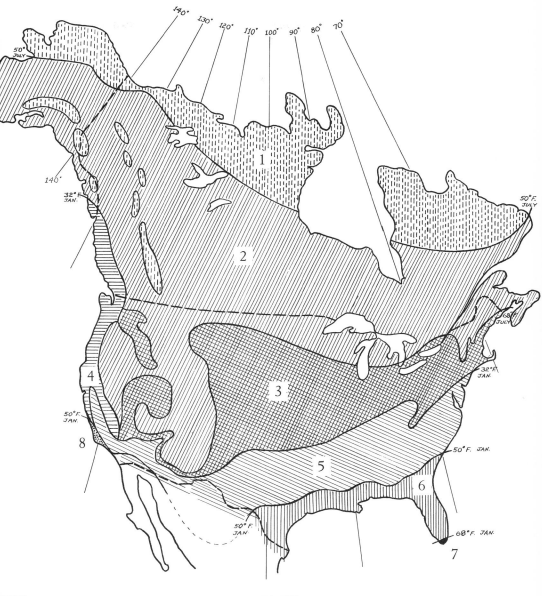

Temperature is mapped by *isotherms*, lines connecting places having the same average temperature for a stated period. If our continent were a rectangular land surface of the same altitude throughout, the isotherms would tend to be parallel east-west lines. But our continent, as we have seen, is far from flat and uniform in outline. In the eastern United States lower temperatures prevail in the higher elevations of the Appalachian Highlands, and as a result the isotherms dip southward in that region. The Atlantic Ocean, tempering the coastal climate, deflects the isotherms northward along the seacoast. In the western United States also, low temperatures in the high mountain ranges cause the isotherms to deviate far to the south, bringing an arctic climate to the high elevations of the Rocky Mountains and the Pacific Coast Ranges. The Pacific Ocean moderates the coastal climate so that isotherms of that region actually extend north and south, rather than east and west.

North America can be divided into a number of temperature belts on the basis of whether the winters are cold, cool, mild, or warm. These belts are temperature zones whose boundaries are isotherms of special ecological significance. Most northern is the 50° F. isotherm for the warmest month (July) that marks the northernmost limit of tree growth. This forms the boundary between a cold winter, cool summer belt and a cold winter, mild summer belt. A second significant isotherm is that of 68° F. for the warmest month. This, the southern limit of the conifer forests, forms the boundary between the cold winter, mild summer and the cold winter, hot summer belt. The latter in turn extends to the 32° F. isotherm for the coldest month (January). Between this isotherm and the 50° F. isotherm for the coldest month lies a cool winter, hot summer temperature belt. The 50° F. temperature average for the coldest month marks the northernmost limits of subtropical wildlife, and hence the northern boundary of a mild winter, hot summer belt. The extreme tip of Florida lies south of the 68° F. isotherm for January; this very small area has an always hot climate, which makes possible an abundance of tropical plants and animals that could not survive in the other temperature belts.

These temperature belts present different extremes as well as different averages of temperatures and so set up conditions in the environment which affect the activities of plants and animals living in these regions. Temperature affects not only the growth, size, appearance, and habits of organisms but also their reproductive activities. Extreme heat, when associated with decreased rainfall, makes life hazardous for species that cannot avoid water loss. With rainfall adequate, the warmer the temperature, the more lush and varied is the vegetation as well as the animal life. Extremely low temperatures and short growing season, on the other hand, create an environment in which only specially adapted plants and animals can survive.

The Climate of North America: Precipitation

The rain and snow that furnish the water so vital for plant and animal life do not fall uniformly over the various physiographic and temperature regions. The Olympic Peninsula of Washington is deluged with 146 inches a year, and southeastern Alaska has an annual precipitation of 150 inches — more than 12 feet of water a year! By contrast, southern portions of the Intermontane Plateau receive less than 2 inches a year. In Death Valley, California, an entire year may pass without a drop of rain. In many mountainous areas the precipitation comes as snow; in the high Sierra Nevada this may measure 300 inches (= 30 inches of rain) a season.

The distribution of precipitation is mapped by *isohyets*, lines connecting places with the same average for a stated period. The isohyets in general extend in a north-south direction rather than the east-west of the isotherms. The rainfall belts therefore cross the temperature belts at right angles, resulting in a gridlike combination of these two important factors of the climate. Certain isohyets are more important ecologically than others, just as we discovered the case to be with isotherms. These significant isohyets can be used to delimit a few major precipitation belts. These are the isohyets denoting an annual rainfall of 40, 20, and 10 inches. Where the annual rainfall is 40 inches or more, a maximum precipitation belt occurs which encourages luxuriant forest growth, with its attendant wildlife. Between the 40- and the 20-inch isohyets lies a region of ample but not abundant precipitation where grasslands begin to replace forests. Since most trees require 20 inches or more of annual rainfall, this isohyet sets the limits for forest growth. A region of moderate rainfall exists between the 20- and the 10-inch isohyets, creating a semi-arid belt in which grasses are the predominant type of vegetation. Where there is less than 10 inches of precipitation a year, the result is an arid region in which plants and animals require special adaptations for survival. Such regions are usually deserts.

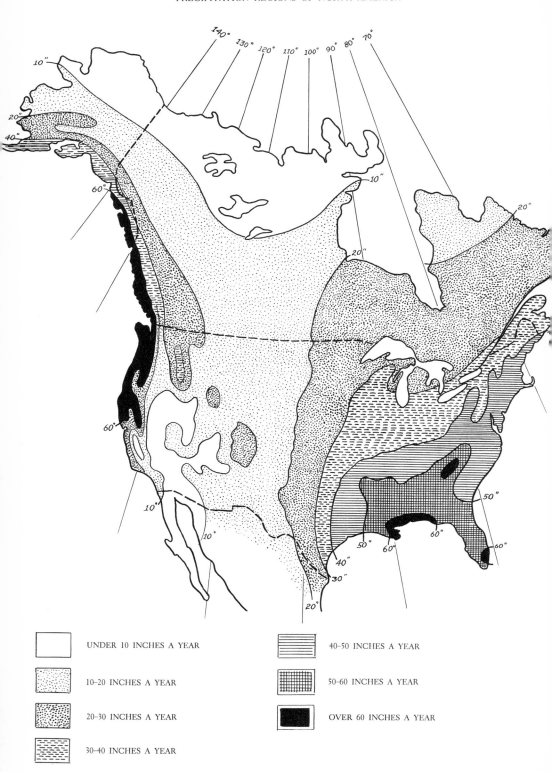

Climate and Life Zones

Temperature and rainfall often go hand-in-hand in their effect on wildlife. They combine to form a small number of *life zones* characterized by uniform climatic conditions and therefore the same types of plants and animals. In 1894 an American biologist, C. Hart Merriam, proposed such zones as a basis for the ecological classification of environments. The boundaries of each life zone are certain isotherms within which live organisms adapted to the prevailing temperature factors. The boundaries also take into consideration isohyets which are of ecological significance. In eastern North America where the temperature belts succeed each other in an orderly sequence from north to south, the resulting life zones also follow a simple horizontal sequence. In western North America, however, the mountains complicate this sequence with variations in altitude replacing those of latitude. Hence the life-zone sequence is a vertical one changing with elevation. As a result the western mountains present a dramatic display of many life zones within a short distance; climbing a thousand feet in altitude equals a horizontal dis-

Life zones form a vertical sequence in the mountains, as represented here by the central Rocky Mountain region.

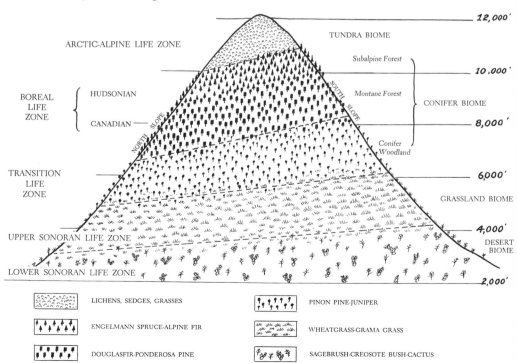

tance of 600 miles at sea level. In many of the southwestern mountains one can travel ecologically a distance of 5000 miles — equivalent to the distance from the Rio Grande to Hudson Bay — while actually traversing only three miles up the mountainside.

There are eight major life zones, each representing a certain combination of temperature and precipitation factors. The *Arctic-Alpine Life Zone* is found in two different areas of North America. One is the coastal belt that fronts on the Arctic Ocean. In this frigid area the growing season is very short; at Point Barrow it consists of only seventeen days. Under such living conditions growth of large plants is inhibited and the tundra growth of lichens, mosses, and grasses predominates. Here almost all the vertebrates are warm-blooded ones, the birds and mammals; and even these are reduced to a few species. The other arctic-alpine area is found on the upper slopes and summits of high mountains, where living conditions duplicate those of the Arctic. This is the life zone above timberline, extending to the elevations where snow and ice form a perpetual ground cover. It too is a region of tundra vegetation and of a few hardy species of birds and mammals.

The *Boreal Life Zone* offers a more temperate climate and less rigorous living conditions. It is sometimes divided into a colder Hudsonian Life Zone and a milder Canadian Life Zone. In eastern North America the Boreal Life Zone is bounded on the south by the 68° F. July isotherm and on the north by the Arctic-Alpine Life Zone. Summer temperatures are higher and winter temperatures less severe than in the Arctic-Alpine Zone; the growing season is longer, being 89 days at Fairbanks and 150 days in northern Maine. Trees can grow in this life zone, and the forests provide shelter for a greater variety of wildlife than in the treeless tundra. The Hudsonian subdivision is predominantly an evergreen forest of spruce and fir, with moose and wolverine as typical mammals. It has the coldest environment in which forests can develop. In the northern lowlands, it merges gradually into the tundra, the trees becoming smaller and more scattered until finally the vegetation is dominated by the nonwoody plants. At high elevations, the Hudsonian zone terminates at timberline.

The *Transition Life Zone*, as the name implies, is a border zone between the northern cold-winter life zones and the southern mild-winter regions. This includes the portion of the Appalachian Highlands which is not in the Canadian Life Zone, the southern Great Lakes area, and the northern part of the Central Lowlands. The growing season is lengthened, being 198 days at Chicago and 211 at New York City. Here

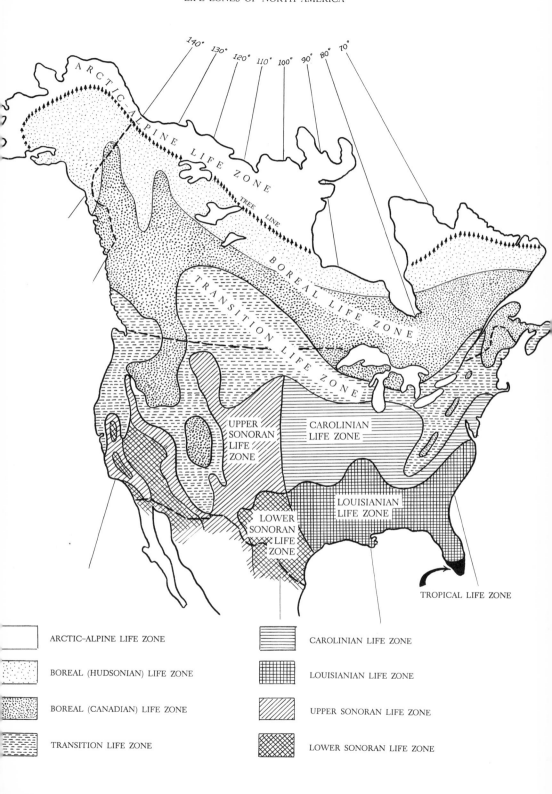

members of the northern evergreen forest mingle with the deciduous oaks and hickories. Formerly this area was almost entirely deciduous forest, but most of it has been removed to provide agricultural land. In the western mountains the Transition Life Zone is a temperate but semi-arid region in which the forests are often made up of stunted piñon pine and juniper.

In the eastern United States, south of the Transition zone, is the milder *Carolinian Life Zone*. It is a region of cool winters and hot summers, in the humid belt where precipitation is 40 inches or more annually over most of the area. The Carolinian zone stretches from the northern part of the Atlantic Coastal Plain to eastern Texas. The growing season has increased to 218 days at Richmond and 238 at Memphis. This too is part of the great eastern deciduous forest, which reaches its maximum development in this life zone. The dominant trees are tulip poplars, holly, magnolia, dogwood, and numerous kinds of birch and maple. The animal life reveals an increasing number of amphibians and reptiles, and a rich and varied population of birds and mammals.

The western equivalent of the Carolinian zone is the *Upper Sonoran Life Zone*, bounded at higher elevations by the Transition zone, on the east by the 20-inch isohyet. It includes the semi-arid western portion of the Central Lowlands, and lower elevations in the Rocky Mountains and Pacific Coast Ranges. In the semiarid region of Nevada, the Reno temperatures and precipitation are typical: a 32° F. January average, and an annual precipitation of eight inches. The vegetation consists chiefly of desert grasses and sagebrush on the drier slopes, willows and aspens in the moister valleys. It is the home of coyote and elk, of lizards and rattlesnakes.

Most of the southeastern United States lies in the *Louisianian Life Zone*. High and dry ground is forested with pine, oak, and palmetto; low swampy ground with cypress. The absence of low winter temperatures and the constantly moist climate encourages the abundance of amphibians and reptiles as well as many birds that have ranged northward from the tropics.

The western counterpart of the Louisianian is the *Lower Sonoran Life Zone*, also a region of moderate temperatures; but it differs in the lessened precipitation. In this zone lie the American deserts, a region of mesquite, creosote bush, and cacti; of numerous snakes and lizards, desert rodents and birds. Some portions of this life zone, as at Yuma, Arizona, have a growing season of over 340 days, but throughout the area rainfall is negligible, usually less than 4 inches and often entirely lacking for a whole year.

The *Tropical Life Zone* is the small but biologically intriguing region south of the 68° F. winter isotherm that crosses the tip of Florida. Here the uniformly high temperatures combined with heavy precipitation and a full-year growing season provide a most favorable environment for abundant plant and animal growth. It is a region into which tropical plants have spread northward to our continent — mangroves, palms, orchids; flamingos, manatees, crocodiles. This life zone represents the hottest extreme of living conditions, and the most southern of our exploration of American wildlife; just as the Arctic-Alpine Life Zone represents the coldest extreme, and the farthest north of our ecological field trips.

We can now retrace our journeys along Route 1 and Route 40 with a better understanding of how the topography, temperature, and rainfall of different geographic areas determine the kinds of plants and animals we encountered. Perhaps now it is obvious that the landscape can best be interpreted in terms of crossing isotherms and isohyets rather than political boundaries; and that life-zone boundaries are far more significant than state lines.

The entire trip along Route 1 takes us through only two physiographic regions, the Appalachian Highlands and the Atlantic Coastal Plain; it is practically a sea-level route. We remain in one rainfall belt — that of 40 inches or more a year. Since this is ample for tree growth, our trip is entirely in a forested area. However, we do pass through several temperature belts and traverse five life zones in our north-south progress. In northern Maine we are at the edge of the Canadian Life Zone, with its spruce-fir forest and northern mammals. In New England, south of Maine, we pass through the Transition Life Zone with an increasing abundance of deciduous trees and a greater variety of

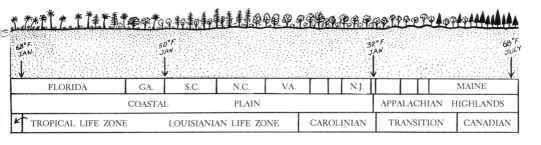

An ecological journey on U.S. Route 1 from Maine to Florida.

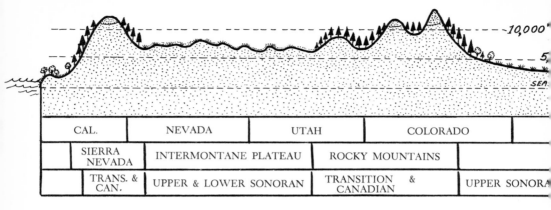

An ecological journey on U.S. Route 40 from the Atlantic to the Pacific Ocean.

birds and mammals. At Baltimore we enter the Carolinian Life Zone, and notice the gradual increase of southern wildlife. The increasing mildness of the Louisianian Life Zone becomes apparent as we cross the 50° F. January isotherm near Savannah, where we see our first palms and discover the greater numbers of amphibians and reptiles. As Route 1 crosses key after key on its way to the tip of Florida, the impact of the Tropical Life Zone with its exotic wildlife brings to an exciting climax our ecological journey along the eastern edge of our continent. Traversing the five life zones has required a five-day trip of some 2000 miles. If we had traveled up the slopes of the Chiricahua Mountains in Arizona we would have visited five equivalent life zones in fifty minutes and less than ten miles!

When we begin our trip westward on Route 40, we are on the coastal plain of Maryland, in the Carolinian Life Zone. The succeeding stages of our journey will take us through a bewildering alternation of life zones, because of the great topographic variety along the route; it will also be a journey marked more by crossing isohyets than isotherms. In the Appalachian Highlands we pass briefly through the Transition Life Zone with its mingling of evergreen and deciduous trees, forming dense forests in the humid climatic belt. We return quickly to the Carolinian Life Zone in Ohio, where the deciduous woodlands thin out and grasses become more and more dominant as we pass through Indiana and Illinois. After leaving the Mississippi River, somewhere in Kansas we cross the significant 20-inch isohyet. Here we bid farewell to woodlands and are surrounded by prairie grasses, the home of prairie dogs today, of herds of bison yesterday.

The highway climbs slowly across the Great Plains, reaching the western mountain ranges in the Upper Sonoran Life Zone of Colorado,

KANSAS	MO.	ILL.	IND.	OHIO	PA.	MARYLAND
NTRAL	LOWLANDS			APPALACHIAN HIGHLANDS		COASTAL PLAIN
CAROLINIAN	LIFE		ZONE		TRANS.	CANADIAN

a land of sagebrush and jackrabbits. The increased elevation at Denver takes us to the western Transition Life Zone, where we again encounter forests of mixed evergreen and deciduous trees, and skirt the lower edge of the Canadian Life Zone with its dense evergreen forests. Taking leave of the Rocky Mountains and its cool moist climate, we descend to the Intermontane Plateau in Nevada, a region of little precipitation where again we are in the Upper Sonoran Life Zone. The altitudinal sequence is repeated as we leave the rain shadow east of the Sierra Nevada and climb the mountain passes to enter California. High above the road we can see the hardy conifer forests of the Boreal Life Zone. Descending to the dry and hot interior valley of California we pass briefly through the Lower Sonoran Life Zone, another arid region where plant and animal life finds existence very precarious. Up over the coastal hills we re-enter the western Transition Life Zone, conscious of the moist and mild air as we reach San Francisco and the Pacific Ocean.

The east-west journey has given us another vivid picture of the combined effects of topography, temperature, and precipitation in determining the wildlife, which changed daily as we proceeded westward. We passed through five life zones: Boreal, Transition, Carolinian, Upper and Lower Sonoran. We had an opportunity to see all the major surface features of the continent from the Atlantic Coastal Plain and Appalachian Highlands, the Central Lowlands, and the Rocky Mountains to the Intermontane Plateau and the Pacific Coast Ranges. Our trip covered three temperature belts: the cold winter, hot summer one of eastern and central United States, the cold winter, mild summer one of the Rocky Mountains and intermontane plateaus, and the cool winter, mild summer belt of coastal California. The westward trip was also a succession of crossing isohyets: the 40-, 30-, 20-, and 10-inch isohyets until we reached California; and then in quick succession a reversal of isohyets from the 10-, 20-, and 30- to the 40-inch isohyet again. Like the journey along Route 1, this trip has shown us that the significant boundaries, to the ecologist, are topographic and climatic ones rather than arbitrary state lines.

The Biomes of North America

Earlier we were approaching our planet from outer space, planning a landing somewhere on North America. From this distance we were able to make out the major features of the continent: the two mountainous rims bordering the Atlantic and Pacific Oceans, the great troughlike interior lowland, the plateaus, and the coastal plain. If we had not been concentrating on the physiography of North America, we would have noticed certain conspicuous aspects of the living mantle which clothed these surface features.

If we are making this landing in summer, when the sun has taken its position north of the equator, we shall see large areas with distinctive colors, indicating regions with different types of vegetation. Along the margin of the Arctic Ocean lies a gray-green, treeless plain. South of this stretches a transcontinental belt of dark green, an unbroken forest with prongs reaching southward on each of the mountainous rims of the continent. The eastern half of the land, south of this forest, is also covered with trees, but these give a more delicate green color to the region, with many open spaces. At the southern tip of the Florida peninsula is a tiny lush-green spot denoting another type of vegetation, an outpost of the tropical lands beyond the continent. The central part of North America is another treeless plain, grass-green in its eastern part, a yellowish green in its western part. The treeless area comes to an abrupt halt at the edge of the Rocky Mountains. Tucked away in the southwestern part of North America lies a sun-baked expanse which seems devoid of life, lying in the rain-shadow of the coastal mountains. Thus the entire continent appears to be divided into a few large vegetation areas. Each of these is a biome.

A biome is a wildlife community of considerable size, often a thousand miles or more in extent and occupying a region in which climatic conditions are relatively uniform. Thus the temperature and precipitation belts determine the boundaries of each biome. The plant and animal life of the earth form six major biomes, all of which can be found in North America. The northernmost portion of the continent fronting on the Arctic Ocean is a gray-green and treeless *tundra* biome.

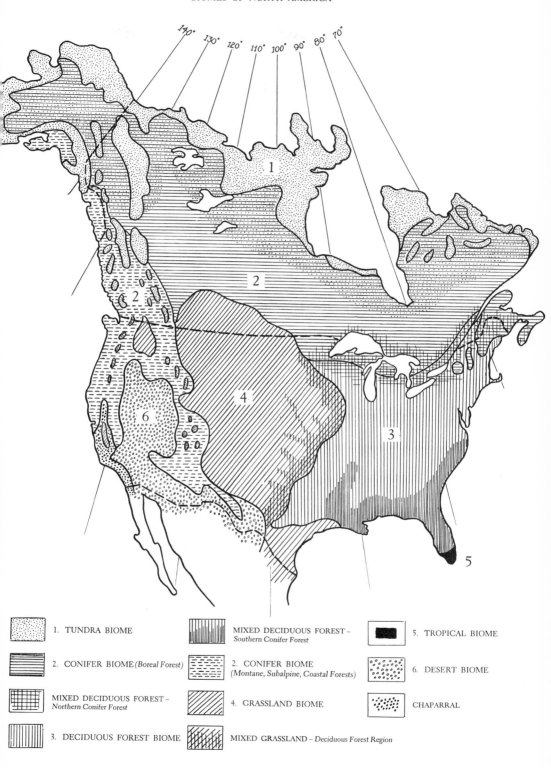

South of the tundra is the dark green belt of the *conifer forest* biome. A lighter green-tree belt, still farther south and extending over the eastern United States, is the *deciduous forest* biome. The interior lowland of North America forms a *grassland biome*, separating the deciduous forest from the western conifer forest. In the southwestern United States, the most arid portion of the continent, lies the *desert biome*. The tip of the Florida peninsula is the only portion of the continent reaching into the *tropical biome*. The boundaries between the biomes are not clearly defined but often include a middle ground where the wildlife of adjacent biomes mingle to form a transition region. This is particularly true where the conifer forest merges with the tundra, and the deciduous forest adjoins the grassland.

With this total picture of the biotic community in mind, we conclude our ecological view of environment and life in North America. Equipped with this knowledge we are ready to explore in greater detail each of the great biomes of our continent, and to appreciate the relation of the physical and living environment of each to its particular types of wildlife communities. From the tundra to the tropics, each biome gives us a concrete idea of a biotic community in action, of life attempting to come to terms with its environment. In each biome we shall see how the ecological view helps to interpret the present wildlife relations that exist in different parts of our country, and see the role of man in modifying these relations.

8 / The Tundra

LIFE REVEALS its most fascinating aspect when it exists in a region where the odds are often against survival. William O. Douglas, in *My Wilderness: East to Katahdin*, was impressed by this fact when he wrote, "The infinite variety of life in the wilderness shows how glorious the products of adversity can be." The tundra biome is a wildlife community living at the frigid frontier of life, where plants and animals have advanced to the very limits of protoplasmic endurance in the polar desert of ice and snow and extremely low temperatures.

The term tundra comes from the Russian *toondra*, meaning a marshy grassland with permanently frozen subsoil. Many forms of life are excluded from this arctic-alpine life zone because they cannot adapt to the low temperatures and lack of protective shelter. As a result the tundra community includes relatively few species in each niche, and the energy flow results in very short food chains. Survival is a competition more with the physical environment than with other plants and animals. The most extensive portion of the tundra biome is a circumpolar region

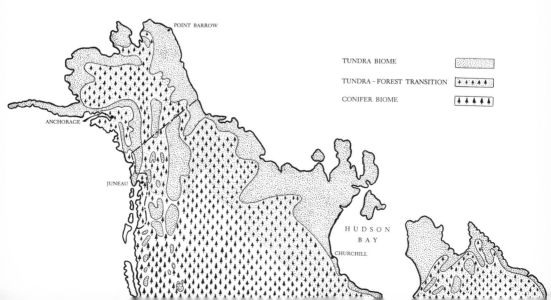

THE TUNDRA BIOME AND ADJACENT CONIFER FOREST

astride the Arctic Circle; this is the arctic tundra. That which exists elsewhere in North America occurs as discontinuous areas above timberline on high mountains, and is known as alpine tundra.

Arctic tundra is found in Alaska and Canada. Its northern boundary is the Arctic Ocean and the Bering Sea; its southern boundary is the limit of tree growth, determined largely by the 50° F. summer isotherm. This extends, as we have seen, from the Labrador Coast to southern Hudson Bay, continues northwestward past Great Slave Lake and Great Bear Lake, and terminates on the western coast of Alaska. This southern boundary of the tundra is not a definite one, but varies with the topography in a transition region where tundra merges with the vanguard of the conifer forests. Hardy trees, chiefly alders and willows, advance northward in sheltered hollows, and tundra grasses and lichens extend southward on exposed ridges. Thus although the arctic tundra is essentially a treeless biome, scattered groups of shrubs and small trees penetrate it along the streams. Tundra is a roadless biome, and as a result inaccessible except by air or by water. At Churchill, Canada, on Hudson Bay or at the terminus of the Alaska-Canada highway at Circle, Alaska, one can catch a glimpse of what this tundra biome is like: a seemingly endless expanse of rolling, grass-and-lichen-covered land dotted with innumerable swamps and ponds.

Alpine tundra is found at high elevations on the summits of the Rocky Mountains, the Cascade Ranges, and the Sierra Nevada. In the eastern United States there is little alpine tundra, and such as there is occurs chiefly on Mount Katahdin in Maine and the higher peaks of the White Mountains in New Hampshire. Today alpine tundra is separated from the arctic tundra, but they were continuous when the ice receded after the last glacial period. After the Ice Age, as the climate became milder and climatic succession took place, forests spread northward and climbed the mountain slopes, leaving isolated tundra of high mountains as islands surrounded by intervening forest communities. Living conditions are much the same, however, in both the arctic and the alpine tundras. As a result their wildlife reveals similar adaptations for existence in an environment where low temperatures and rigorous living conditions prevail most of the year. The outdoorsman who enjoys climbing mountains is able to reach the arctic tundra in many parts of the United States. Perhaps the most visited of such tundra areas is that in the White Mountains which confronts the hiker who has toiled up the face of Tuckerman Ravine on Mount Washington and reaches the alpine meadows above timberline. Here he sees the struggle for

Tundra with willow-shrub vegetation. Arapaho National Forest, Colorado. (*Credit: U.S. Forest Service*)

survival which the grasses, mosses, and lichens carry on amid a jumble of granite boulders, in a miniature tundra biome that extends to the summit.

The Physical Environment of the Tundra

Arctic tundra consists for the most part of low rolling plains with rounded ridges and shallow depressions where poor drainage creates many bogs, streams, and lakes. Continuous frost action molds the landscape into a pattern of mounds and hummocks as rock fragments are shifted by alternate freezing and thawing. The surface soil settles into the hollows, carrying with it clumps of lichen and moss and leaving the ridges bare. Much of the substratum is therefore unstable: wet, soggy, and acid in the depressions and barren on the exposed higher ground.

The winters are long and cold, the summers short and cool. Winter temperatures remain at zero or below for many months of the year,

Alpine tundra. Rio Grande National Forest, Colorado. (*Credit: U.S. Forest Service*)

summer temperatures average less than 50° F. The short summer season places a premium upon rapid production of seeds and quickly maturing young. The continuous low temperatures prevent complete thawing of the ground and result in permafrost that often extends to a depth of several hundred feet. Only the upper foot or two of ground thaws during the short summer, and the soil so provided is too shallow and poorly drained for the root growth needed by most trees.

Annual precipitation is less than ten inches, the greatest amount occurring as summer rain; snowfall is relatively scant. Light in the arctic tundra is another critical factor because of its limited amount and low intensity. Daylight hours are extremely few for part of the year, and even in the longer summer days the light is weak because of the low angle at which the sun's rays strike that part of the earth. Such light conditions reduce the energy input of the ecosystem, lowering the productivity of green plants and thus limiting the amount of life the community can support. Other physical factors are storms and high winds. Combined with the low temperatures and scarcity of water in the winter months, they limit the size of plants and restrict the amount of forage for the herbivores.

Alpine tundra shares with the arctic tundra many of these physical conditions: low temperatures, high winds, and unstable substratum. But alpine tundra differs in having greater precipitation, mostly in the form of snow. This abundance of water, nevertheless, is of little value to the vegetation, since most of it disappears as runoff in the loose rocky soil. The rocky terrain with its limited amount of soil offers precarious footing for rooted plants. Another significant factor here is low barometric pressure combined with reduced oxygen content of the atmosphere. This requires special physiological adaptations on the part of the animal population.

The Living Environment: The Plant Life

The base of the pyramid of life in the tundra consists, as in all biomes, of the food producers. Here they are the plants adapted for living on rocks or sterile shallow soil and able to survive low temperatures and reduced water intake for most of the year. Plants needing no soil are numerous in the barren portions of the tundra; they are the lichens and mosses that initiate the biotic succession transforming bare rock into soil. Of the seed plants in later successional stages, annuals are scarce, being at a disadvantage in the short growing season during which they would find it difficult to germinate, reach maturity, and produce seeds. Far more numerous are the herbaceous perennials. These are the grasses and sedges as well as the wildflowers like lupine, goldenrod, saxifrage, and anemone which all burst into bloom during the short summer. Some woody perennials are also present, many of them evergreen shrubs that possess special adaptations for tundra life. The few hardy trees are deciduous birches, willows, and alders; of the conifers an occasional white spruce or black spruce stands as a lone outpost of the boreal life zone to the south.

Four types of smaller plant communities make up the tundra biome: lichen-moss, grass-sedge, bush-heath, and scrub-birch. The *lichen-moss community* is the pioneer community, colonizing bare rock surfaces with gray, black, or yellow crust lichens. These soon become overgrown with hardy brown rock mosses and gray-green cushions of reindeer lichen and Iceland "moss," also a lichen. The wet hollows become filled with sphagnum, or peat moss, whose spongy carpet often forms the base of hummocks or reindeer lichen. These are the dominant plants of the exposed tundra, the main diet of caribou and emergency rations for many other tundra mammals.

The *grass-sedge community* develops on the deeper and richer soil provided by the lichen-moss pioneers. This is often widespread, giving many areas of the tundra a prairie-like aspect. The shallow but spreading root system of these perennials remains alive during the winter, producing new green shoots rapidly with the early spring thaws. Grasses are a staple item in the diet of all the grazing animals, in addition to being the chief food of rodents and hares. Grass seeds are prized by some of the smaller mammals and by many of the tundra birds. The frozen but edible aboveground parts of grasses are winter forage for the larger herbivores, the caribou and muskoxen.

The *bush-heath community* is often the climax vegetation of the biome. Seedlings of woody plants establish themselves among the matted masses of lichens, mosses, and grasses, taking advantage of the enriched soil beneath. Even when full-grown, these shrubs are stunted and form prostrate mats or compact cushions among the rocks and other vegetation. They include a number of evergreens and many members of the heath family: Labrador tea, alpine azalea, bilberry (a species of blueberry), mountain cranberry, and bearberry. The berries from these shrubs add a welcome variety to the restricted menu of the tundra herbivores.

The only trees of the tundra are found in the *scrub-birch community*; many are no larger than shrubs. The birches are diminutive cousins of the larger species in the boreal life zone. The willows provide forage in all seasons of the year by their foliage, catkins, stems, branches, and roots. Some alders reach a height of eight feet, the greatest stature of any tundra tree. These hardy species form thickets along stream margins where they are protected from the wind and where the ground thaws deeper, permitting more root growth than on the upland tundra.

Similar communities develop in the alpine tundra. In this high region bordering the snow-capped summits, the lichen-moss community occupies the most exposed habitats below the summit. This community is made up chiefly of crust lichens and rock mosses; fewer cushions of reindeer lichen and sphagnum moss are seen than in the arctic tundra. At lower elevations the grass-sedge community forms a conspicuous fringe adjacent to the bush-heath vegetation of the timberline. The evergreen heaths mingle with the grasses and the advance guard of the timberline trees. The elevations at which these alpine tundra communities are found vary with the latitude of the mountains. On such

Timberline on Cove Mountain, Utah, Manti National Forest. The stunted "elfin timber" consists of Engelmann spruce and Rocky Mountain white pine. (*Credit: U. S. Forest Service*)

eastern summits as Katahdin and Washington, alpine tundra begins at 3000 to 4000 feet, but in the Rocky Mountains its lower limits may be from 8000 to 10,000 feet. At timberline the spruces, firs, birches, and alders form dwarfed, prostrate growths of twisted and streamlined trees often referred to as "elfin timber." Here trees one hundred years old may be only a few feet in height, but forming such dense mats at the edge of the tundra that one can walk on top of the "forest" as if it were a meadow.

Most of the adaptations of tundra plants reduce water loss by transpiration; they are remarkably like those of desert vegetation. The growth form is compact and spherical, a shape that has as much photosynthetic volume as possible, with a minimum of surface exposed to the elements. This hemispherical and cushionlike form serves other purposes also. It enables the plant to hoard castoff organic debris on the ground beneath and so add to its limited soil supply; it also prevents evaporation of water from the ground immediately below the plant. The leaves of tundra plants are generally small and leathery, with a waxy or hairy surface, or with an inrolled margin. These are all adaptations to lessen transpiration.

LEMMING

The Living Environment: The Herbivores

Another niche in the tundra community is that of the herbivores. Some have been tempted to come into the tundra during the summer from the wooded biomes to the south. Among these adventurers are ground squirrels, mice, voles, and hares. Others are native tundra inhabitants with adaptations to fit them for permanent residence. These include the lemmings, arctic hare, barren ground caribou, and muskox.

The most abundant tundra animal is the LEMMING, an important link in the food chain leading from vegetation to the carnivores. Lemmings are small rodents with enormous appetites, consuming great quantities of grasses, sedges, berries, and willows. The common brown lemming, found also in the boreal forest, retains its grayish-brown coat throughout the year; the collared lemming, restricted to the tundra, changes to a protective white coat in winter. Lemmings do not hibernate, but remain active during the entire year. Their tunnels and burrows are everywhere, leading by subterranean passageways to winter feeding areas of buried mosses and roots. Lemmings breed at a phenomenal rate, with a half-dozen young in each litter and several litters a year. Such reproductive enthusiasm leads to periodic population explosions that occur every three to five years. At these times, overcrowding of their habitat and the resulting scarcity of food set off spectacular mass migrations. During these peak years the tundra swarms with the tiny rodents, which provide a feast for the many feathered and furry carnivores and bring about a corresponding increase in the populations at higher levels in the life pyramid.

Tundra is the home of the ARCTIC HARE, largest of all American hares; a full-grown adult weighs from ten to twelve pounds. In summer the fur is grayish brown, with white underparts; but in winter the hare dons an immaculate white fur coat to be seen in the tundra. In spite of its helpless appearance, this hare is well able to take care of itself. It can travel over snow-covered ground with ease and has an uncanny ability to dodge and change its course when pursued. It is a hardy animal that seldom resorts to hiding in burrows. Instead, it weathers the winter storms crouched on the snow, letting the drifts pile up around it. Arctic hares often congregate, motionless on the snow, relying on their camouflage for protection. In summer they feed on grasses, crowberry, and other small seed plants. In winter they dig through the snow with their strong front paws to reach buried grasses and willow twigs.

BARREN GROUND CARIBOU are the most abundant of the large tundra herbivores. They are large, heavily built deer, weighing as much as 400 pounds. Both sexes bear antlers, which reach a spread of four feet. Caribou fur is grayish brown; a white neck ruff forms a "beard" on the upper chest, and white fur may extend along the flanks to the hips. Caribou living in the northern part of the tundra are lighter in color than those living elsewhere, some varieties being almost pure white. The fur coat affords excellent protection against the cold; among the short hairs of the underfur grow long quill-like guard hairs to insulate the body. This air-filled covering also gives the caribou added buoyancy as it swims across streams and lakes. Another adaptation of the caribou is unusually large, broad feet; the two hoofs spread wide, leaving a hollow center with a sharp raised outer rim. Such hoofs provide support on both snow and spongy lichen beds, and nonskid traction on ice. In summer, barren ground caribou feed on willow and birch foliage, grasses, and sedges; in winter their diet consists chiefly of lichens. The caribou of the tundra are often plagued by myriads of bloodsucking insects which swarm around the ponds and streams in early summer. Flies and mosquitoes frequently force these and other large herbivores to resort to windswept ridges or take refuge in the water to escape the plague of insects.

The annual migration of thousands of these great herbivores between their summer and winter feeding grounds remains one of the wildlife spectacles of the American continent. Such herds may stretch for miles

BARREN GROUND CARIBOU

in length and spread out to a mile-wide front. When they reach their southern range, along the edge of the boreal life zone, the barren ground caribou mingle with their close relatives the woodland caribou, a species of the evergreen forest to the south of the tundra. The arch enemy of the caribou herds is the gray wolf; in the open, however, caribou can outdistance this predator. Barren ground caribou are still fairly abundant, especially in Alaska and Yukon Territory, where there are an estimated 2,000,000 animals. A major threat to their survival is destruction of their lichen feeding grounds by fire or overgrazing. Lichens grow slowly, and an area burned or otherwise disturbed may take thirty years or more to recover. Reindeer are the barren ground caribou of the Old World which have been domesticated. Reindeer have been introduced into Alaska as an aid to sustaining the Eskimo population.

Symbolic of the tundra wildlife community is the MUSKOX, so named because of its oxlike appearance and the musky secretion released from glands beneath the eyes. Muskox is a short-legged, heavily built member of the cattle family; it resembles a small white-socked bison. Bulls have a shoulder height to five feet and a maximum weight of 900 pounds. Both sexes bear broad horns that almost meet in the middle of the forehead, then sweep outward and downward in a graceful curve, and end in a sharp and effective upturned tip. No other mammal is so adequately clothed to withstand the arctic climate. The hair, over two feet long on the chest and sides, hangs like a skirt around the animal. In addition there is a warm suit of thick woolen underwear. With such protection, muskoxen rarely seek shelter even in the worst winter storms, and so have no need to migrate. Like the caribou, they are adapted to travel over snow and ice by broad spreading hoofs, the sharp edges of which grip the frozen surface and provide excellent traction. They feed mainly on grasses and sedges, adding foliage of alders and willows when available. At the first sign of winter the herds of muskoxen desert the willow thickets and grassy lowlands for higher ground, where the surface is usually swept clear of snow and they can dig out frozen bits of vegetation. Being gregarious animals, they are usually found in groups of a dozen or more. When in danger they defend their calves by forming a circle around them; the old bulls on the circumference all head outward, presenting an array of horns which discourages the attack of wolf packs. This maneuver served them well against animal predators, but Eskimos and explorers equipped with rifles could easily slaughter an entire herd when it assumed even this compact formation. Muskoxen have been important animals in the life of the Eskimo, providing meat, fat for fuel, skins and fur for clothing and bedding, horns for making implements.

In contrast to the abundant caribou, the muskox is today one of the rarest of the tundra mammals. It is a survivor of a once widespread race inhabiting the circumpolar regions, a contemporary of the now extinct mammoth and the woolly rhinoceros. At present only a few thousand muskoxen survive. These live in northern Canada between Hudson Bay and Alaska. Prior to 1865 they were found also in Alaska. In 1930 they were re-introduced by bringing a small herd the 14,000 miles from their native home in Greenland and establishing them on Nunivak Island.

Among the birds, the most hardy herbivore of the community is the ROCK PTARMIGAN. It is a northern member of the grouse family, brown in summer but protected in winter by white plumage. In winter, too, the legs and feet become heavily feathered. The diet of the rock ptarmigan consists of buds, twigs, berries, and mosses. It nests in the open tundra, usually in the shelter of a rock or cliff. Its rabbitlike, sharply clawed feet can dig through winter snow to reach buried vegetation. Ptarmigans are a link in the food chain that terminates in the predatory gyrfalcon.

In alpine tundra the herbivore life is not as varied or as abundant as in arctic tundra. Few mammals and birds can find year-round food and shelter in such exposed and rugged environment. Only a rare vertebrate takes up residence in the alpine tundra of the eastern mountains, and these are summer visitors from below timberline. The alpine tundra of the western mountains also has some summer migrants that wander up from the boreal life zone. Such are wapiti, mule deer, and woodland caribou. With them come such carnivores as weasels, coyotes, wolverines, bears, and mountain lions.

The western alpine tundra, however, has a few characteristic herbivores; these are the pika, mountain goat, bighorn and Dall sheep. The rockslides of the western mountains offer little to encourage animal residents. Yet the PIKA, a hardy relative of the hare family, lives among

ROCK PTARMIGAN

the lichen-incrusted boulders and the scant patches of grass. Its hind limbs are not much longer than the forelimbs; short broad ears and inconspicuous tail give it a resemblance to a small guinea pig. The pika's body is clothed in grayish or buff fur that blends with the rocky habitats above timberline. The soles of its feet are covered with hair, which gives firmer footing on bare rocks. The pika spends its entire life more than two miles above sea level, seeming to enjoy the rarefied atmosphere and the chilly windswept tundra. In fact, in the summer it retires to the cool recesses of its home in the rocks. Summer and winter, pikas feed on grasses and sedges. They are industrious animals, preparing for winter by storing piles of well-cured grasses. These drying haystacks can be seen near the entrance to their burrows. Pikas are safe from martens and weasels as long as they stay in their alpine habitat but fall prey to these carnivores when they venture into the wooded areas below timberline.

Another inhabitant of the rocky mountain slopes is the MOUNTAIN GOAT, which is thoroughly at home above timberline in the Rocky Mountains from Alaska to Idaho. This goat is actually an antelope related to the European chamois. It has a white or yellowish fur coat of shaggy fur, retained throughout the year; beneath is a thick woollier coat, which gives added protection against the cold. Full-grown rams measure three feet high at the shoulders and weigh as much as 300 pounds. Both sexes bear slender horns that have a slightly reverse tilt. Mountain goats feed on grasses and sedges in summer and browse in winter on exposed foliage and branches of trees. When other food is unavailable, they eat lichens and mosses. These alpine animals are famed for their climbing feats, which enable them to escape predator enemies — all but rifle-equipped hunters. Mountain goats travel with a stiff-legged and deliberate gait over dangerous trails made less hazardous by their specially constructed hoofs. The sole of each foot is concave to act as a suction cup when forced against a rock. Additional advan-

MOUNTAIN GOAT

tage comes from the fact that the cleft between the two hoofs opens toward the front so that the toes spread apart when the goat descends a steep slope. Although safe in a mountain habitat, when they descend into the wooded valleys in search of food, these mountaineers fall prey to mountain lions and lynx. Mountain goats survive in considerable number because of their inaccessible habitats and, fortunately for them, the low trophy value of their unimpressive horns.

BIGHORN and DALL SHEEP live above timberline in the Rocky Mountains; the bighorn ranges from the Canadian border to Mexico. Dall sheep lives farther north, from British Columbia to Alaska. The bighorn has pale buff to dark brown fur, the Dall sheep is yellowish white in color. Both have massive horns curling forward in a complete circle; those of the ram often measure four feet around the outer curve and are larger than those of the ewe. These wild sheep can climb incredibly steep slopes, a feat possible because of hoofs that provide a suction grip, as do those of the mountain goat. They are also champion jumpers, able to clear a four-foot hurdle and cover a distance of sixteen feet. The food of both species of sheep consists of all available herbaceous plants, shrubs, and foliage of the few alpine tundra trees. In winter they paw through the snow in search of grass. Some migration occurs between summer and winter feeding grounds, a shift either into the woods at a lower altitude, or at the same altitude to exposed ridges where the wind has kept the ground clear of snow. As happens to many alpine herbivores, when in the lowlands they fall prey to wolves, mountain lions, and coyotes.

The Living Community: The Carnivores

In the tundra, as in other life zones, the carnivore is the final link in the food chains of the biome. Cold-blooded vertebrate carnivores are unable to maintain the body temperature necessary for living in the tundra. So they are rare in this biome. Only a few amphibians are found here. One is the Hudson Bay toad, a variety of the common eastern toad, the American toad. Another is the wood frog, which breeds in the shallow tundra pools; this amphibian has the most northern range of any cold-blooded vertebrate. Of the reptiles, only one has been reported — a garter snake found near Hudson Bay. Thus the dominant carnivores of the tundra are mammals and birds. Some are species from the boreal forest which during the summer venture into the tundra. Such are weasels, wolverines, bears, lynx, and mountain lion, which we shall meet in the following chapter. Others are perma-

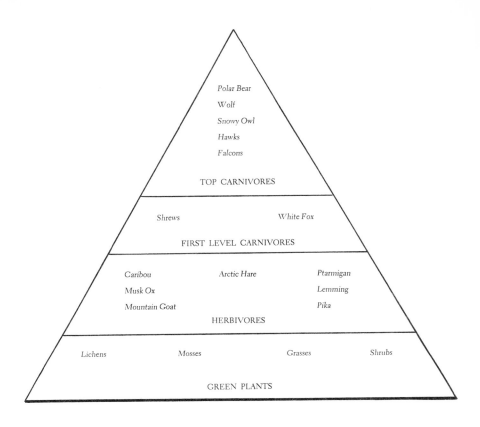

THE PYRAMID OF LIFE IN THE TUNDRA

nent residents: the white fox, gray wolf, and polar bear.

A circumpolar relative of the common red fox is the WHITE FOX, or arctic fox, less than two feet in length and weighing only ten or fifteen pounds. It has a stubby profile, short rounded ears, and bushy tail. The dense fur is grayish brown in summer, but becomes white in winter. It is a hardy carnivore, unmindful of even severe winter weather; during snowstorms it burrows into the nearest drift to wait until safe to venture abroad again. The white fox is primarily an inhabitant of coastal islands. On land its staple diet consists of lemmings. When these small herbivores decrease in number, white foxes migrate southward into the forest edge for food. They eat ptarmigan eggs, spawning salmon, and occasionally plant food. In winter many of the foxes remain on the ice

FOOT ADAPTATION, WHITE FOX

floes, trailing polar bears and feeding on the leftovers from the larger carnivores' meals. In the tundra, the chief enemies of the white fox are bears and snowy owls.

Roaming the tundra in search of smaller animals, prowls the GRAY WOLF, a large predator with a length of almost six feet and an average weight of seventy-five pounds. Its fur is usually pale gray, but one variety has a yellowish-white coat, and it can also be nearly black. This predator is not a fast runner, yet it has tremendous endurance. Most often it must be content with catching ground squirrels, hares, and mice, but when hunting in packs wolves seek out larger prey such as muskoxen and caribou. In pursuing these large herbivores, wolves rarely attack a herd. Rather, they follow it day after day until a crippled or aged individual lags behind and becomes easy prey. Primarily a member of the forest biome to the south, the wolf has retreated to the wilderness of the tundra to avoid extermination by hunters. Today the gray wolf survives chiefly in national parks and protected refuges of Alaska and Canada.

Today the gray wolf survives chiefly in national parks and protected refuges of Alaska and Canada. (*Credit: New York Zoological Society*)

Polar bears, photographed from the air on the shore of Hudson Bay. (*Credit: Ontario Dep't of Lands and Forests, Canada*)

No picture of the tundra would be complete without that symbol of the icebound Arctic, the POLAR BEAR. This circumpolar animal spends most of its active life on ice floes, rarely setting foot on land. In winter it goes ashore, but only for a long deep sleep buried in the snow, where it remains until the following spring. The polar bear is the most amphibious of all the large carnivores, being an expert diver and swimmer. Yet it can also travel over ice and snow at twenty-five miles per hour. It has a long supple neck, small head, and a thick yellowish-white fur coat. A full-grown individual stands four feet high at the shoulders, is almost eight feet in length, and weighs up to 1100 pounds. This fearless and dangerous adversary is truly one of the great carnivores of the world. It preys chiefly upon seals, walrus, stranded whales, and other marine life. It is a top carnivore, the last link in a food chain that has led from tiny shrimplike krill of the sea through intermediate links of fishes and seals to this mammoth predator.

Two birds of prey are also tundra carnivores: the snowy owl and the gyrfalcon. The SNOWY OWL is a barred white bird whose domain is the hilly tundra where it can find elevations on which to perch and scan the surroundings for a meal of lemming, hare, or ptarmigan. In winter when the lemming population reaches its periodic low point, snowy owls often visit the boreal life zone to the south. Such an influx of snowy owls into northern United States occurs about every four years. The GYRFALCON, slightly smaller than the snowy owl, varies in color from white to black. It also patrols the tundra in search of lemmings

SNOWY OWL

and hares; at some months of the year its special prey is ptarmigan. The related peregrine falcon breeds in the tundra but regularly migrates to southern biomes in winter. It preys almost entirely on other birds captured in midair.

The adaptations enabling many animals to live in the tundra include white coloration, thick fur coats, and special modifications of the feet which facilitate locomotion in arctic or alpine habitats. White fur or feathers is of decided protective value, both to herbivores on a snowy landscape with little shelter and to carnivores stalking their prey. Adequate protection against cold and storms is secured by long shaggy hair, as in the muskox, or by double fur coats, as in the mountain goat. Suction-cup hoofs aid locomotion over the terrain encountered in the tundra, providing sure footing on steep slopes to mountain goats and bighorn sheep. The broad-spreading hoofs of caribou and muskoxen enable them to traverse snow and spongy mosses, and such hairy soles of the feet as those of the pika give these animals sure grip on rocks and snow. For all tundra animals, obtaining food in winter is a prime problem. Vegetation buried under snow can be reached by herbivores only by clawing through the snow, and a scarcity of prey at this season demands keen eyes and stamina on the part of the carnivores.

Summer Bird Life of the Arctic Tundra

Marshes, streams, and ponds in abundance become a summer paradise in the tundra for the many species of land and water birds which annually migrate to this biome to breed. Here they find vast unmolested breeding sites and ample supply of insect and plant food. The incredible wealth of this summer bird life can best be described by the British ornithologist James Fisher, who gives a firsthand account of a June exploration of western Alaskan tundra.*

"We looked over an eternal expanse of green and brown: grass, creeping willow, crowberry, bearberry, cornel, and Labrador tea," he writes. "It was a glorious garden of arctic plants . . . never for a moment were we out of sight or hearing of crane, goose, duck, or wader. Everywhere Lapland longspurs, in full black-throated breeding dress, dropped their pretty notes." He goes on to describe waders of a dozen sorts nesting, singing, and scolding on every hand. "The air was full of the lovely rippling trill of the dunlin," he says, and "along the edges of the tarns the little western sandpipers fluttered and twittered." Northern phalaropes were "spinning on the dark mirror of every pool." On their way downriver they ran a "gantlet of waterfowl," Fisher remarks, "such as I had encountered but once before in my life."

According to Fisher, the tundras between Chevak and the coast support about 130 nesting waterfowl to the square mile. Among those he mentions are American scoters, scaup, pintails, lesser Canada and cackling geese, white-fronted geese, oldsquaws, and lesser numbers of swans, mallards, baldpate, teal, and others.

MAN AND THE TUNDRA BIOME

Because of the climate and present unsuitability for exploitation by the white man, the arctic tundra has been altered the least of all the biomes. It still remains a vast unspoiled wilderness, one of the last remaining wildlife sanctuaries of large area on our continent. The use of the land by the Eskimo inhabitants in former times had small adverse effect on the native vegetation and animal life. Trapping and killing of tundra mammals with primitive weapons disturbed the ecological balance very little. But, with the importation of firearms and the white man's civilization, great inroads were made on some species, like the muskox.

In more recent years, human intervention has added two disturbing ecological factors. The introduction of reindeer into Alaska has brought with it the threat of diseases that can spread from the domesticated to the native races of caribou. Potentially far more dangerous has been

* See the chapter entitled "Tundra of the Emperors" in *Wild America* by Roger Tory Peterson and James Fisher (Boston: Houghton Mifflin Company, 1955).

the effect upon the tundra food chains of radioactive fallout resulting from atmospheric bomb tests. The tundra biome, where the dominant vegetation consists of lichens, lies in the path of much of this fallout. Lichens are air plants and as such take into their bodies great amounts of the radioactive material settling onto the tundra. These plants, as we have seen, are an important food for caribou, and caribou is a source of meat for Eskimos. It is only within the last few years that any research has been carried on dealing with this aspect of the tundra ecosystem. Much more needs to be done before we can know whether or not the tundra biome and its inhabitants will be safe from this modern threat which civilization brings to a biotic community.

9 / The Conifer Forest

To THE TRAVELER along the trans-Canada highway or the Alcan Highway leading from Washington to Alaska, the vast conifer forests are a constant companion for thousands of miles. The conifer forests, to the vacation-minded American family, are a cool and refreshing wilderness which forms a setting for our best-known national and state parks. Whether you are exploring Baxter State Park in Maine, the White Mountain National Forest in New Hampshire, the Rocky Mountain National Park of Colorado, or Yosemite National Park in California, you are in a wildlife community dominated by the evergreen conifers.

The conifer, or evergreen, forest is one of the largest biomes of North America. It lies between the 50° F. and the 68° F. summer isotherms and so occupies the large part of Canada and Alaska which lies in the boreal life zone. Here the biome forms a continuous forest that merges northward with the arctic tundra. To the south the conifers mingle with the angiosperm trees of the deciduous forest biome. Southward extensions of the conifer forest reach into the Appalachian Highlands and the western mountains, where the boundaries of the forest are determined by altitude. At high elevations the conifers merge with the grasses and lichens at timberline; at its lower limits, the conifer forest gives way to deciduous woods and grasslands.

THE PHYSICAL ENVIRONMENT OF THE CONIFER FORESTS

The physical environment of this biome is not as inhospitable as that of the tundra, yet only hardy and well-adapted plants and animals can be permanent residents of the community. In Canada the forest occupies the granite dome of the Laurentian Plateau, a barren and poorly drained region where the fertile topsoil was removed by successive movements of the continental glaciers of bygone eras. In the United States the conifer forest covers the higher slopes of the mountains, where the topography is rugged and the soil is often rocky and shallow.

Most of the conifer forest is located in the cold winter, cool summer and cold winter, mild summer climatic belts. Here the climate is rigor-

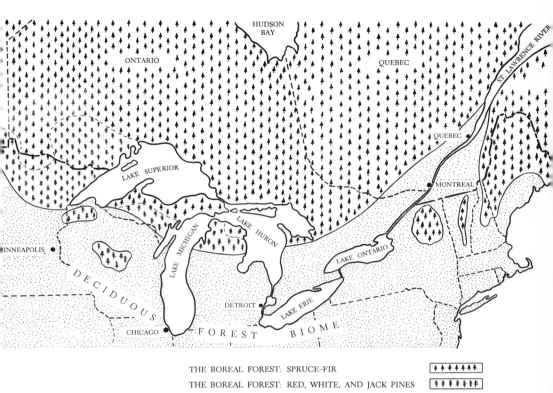

THE BOREAL FOREST: SPRUCE-FIR
THE BOREAL FOREST: RED, WHITE, AND JACK PINES

ous, with long cold winters and short summers. The temperature ranges from a winter average of −20° F. to a summer average of 70° F. The growing season is longer than that of the tundra, being increased to a maximum of 150 days. The conifer forest of the Pacific Coast lies in the always mild climatic belt, and conditions are therefore more equable and moderate, with a winter average of 35° F., a summer average of 65° F.

The entire region lies in the maximum rainfall belt. On the Pacific Coast the forest receives more than fifty inches of rain a year; the annual precipitation elsewhere in the biome ranges from twenty to forty inches. On the western slopes of the Sierra Nevada, the lower conifer forests can thrive with only ten inches of rainfall annually.

The Living Environment: The Plant Life

The green plants which are the primary energy converters for this biome are mainly woody perennials. Of these the dominant species are conifers, well adapted by foliage and growth habits for living in the boreal life zone, as we have seen in Chapter 7. Conifers provide shelter for the smaller plants and for the forest-dwelling animals; their seeds are a food for the herbivorous mammals and the birds. The biome is not entirely a coniferous forest. Some deciduous trees are present; these are for the most part members of the catkin-bearing birch, willow, and poplar families. Although a minority group, the deciduous trees are important in providing a change of diet for herbivores like moose, deer, and rabbits, which feed on their foliage and stems. Other smaller food producers include blueberry, raspberry, and other berry-forming shrubs whose fruits are eaten by herbivores and omnivores alike. The foliage of herbaceous perennials such as grasses furnish additional food for rodents, rabbits, and deer. The dense canopy of the conifer forest allows little sunshine to filter through to the forest floor. As a result the diffused light and high humidity provide an ideal environment for a luxurious ground cover of mosses, lichens, ferns, and herbaceous flowering plants. These types of green plants occur throughout the biome, but they are represented by different species in different areas. Each has its characteristic kinds of trees and undergrowth, occupying the producer niches.

The conifer-forest biome covers such a large area with diversified topography that it has become differentiated into four subcommunities or associations, based on the trees dominating the producer niche. A *boreal forest* occupies all of Canada south of the tundra except for those portions of Alberta and Saskatchewan which are part of the grassland biome. The boreal forest also reaches into the interior of Alaska, and southward into the Great Lakes and New England states. A *subalpine forest* forms a high-altitude conifer zone in the Rocky Mountains and Pacific Coast Ranges, with its uppermost limits at the edge of the alpine tundra. The boundary between forest and tundra often forms a clear-cut timberline. At its lower limits this forest merges into a *montane forest* that extends to the base of the western mountains, where it joins the grassland biome. The fourth association is a *coastal forest* that clothes the mountains along the Pacific Coast and extends to the edge of the ocean.

Boreal Forest. If we should be making our way on foot from Hudson Bay tundra southward to the Appalachain Highlands, we would begin our trip amid a stunted conifer forest stretching from horizon to horizon in an endless community of spruce and fir. The patches of forest are separated by many ponds, swamps, and streams forming the vegetation known as muskeg. As we progress southward the trees become taller and form a more dense forest. In the Great Lakes region, groves of pines overtop the spruces. And in the northern Appalachian

The boreal forest extends over much of Canada south of the tundra. Hudson Bay area, Ontario, Canada. (*Credit: Ontario Dep't of Lands and Forests, Canada*)

Highlands we find the conifers marching up the slopes of the Adirondacks, the White Mountains, and Mount Katahdin. If we continue southeastward we find the spruce-fir forests perched on the rocky headlands at the edge of the Atlantic Ocean, in northern Maine.

The most abundant and typical trees of the boreal forest are the spruce sand firs. Widely distributed is the WHITE SPRUCE, a conifer of well-drained moist soil. Transcontinental in range, it is a prominent member of the community from Alaska to northern New England. White spruces a hundred feet tall grow in British Columbia, but most trees are under seventy-five feet in height. Its foliage is a distinctive bluish green. RED SPRUCE is a species with more southern range, being more common south of the Canadian border. It is a large spruce, growing to a height of one hundred feet, with yellowish-green foliage. Its green spires form the characteristic skyline of northern New England

The white spruces of the boreal forest march to the edge of the Atlantic Ocean in Acadia National Park, Maine.

RED SPRUCE

forests, and climb to timberline in the higher mountains. Farther south red spruces form dark green caps to the ridges of the Great Smoky Mountains, at elevations above 5000 feet. The boreal forest includes scattered stands of BALSAM FIR, a symmetrical and fragrant eastern conifer. It thrives in cool moist habitats of southeastern Canada and northern New England. Seedling balsam firs tolerate dense shade and so often appear beneath the canopy of overshadowing spruces. Balsam fir accompanies spruces to timberline in the northern Appalachian Highlands.

The undisputed monarch of the conifers found in the boreal forest is the WHITE PINE, which attains its best growth in deep sandy loam

The boreal forest in New England includes stately groves of red and white pines. Lake Chocorua, New Hampshire.

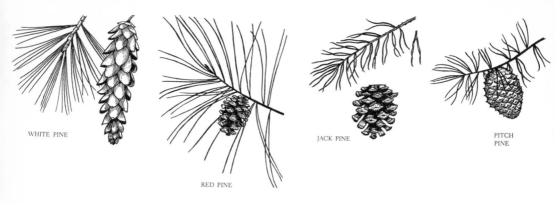
WHITE PINE RED PINE JACK PINE PITCH PINE

with adequate moisture. White-pine forests were formerly abundant in the Great Lakes states and New England. In the Colonial days it proved such a valuable timber tree that few groves have survived two centuries of logging. Giant white pines, several hundred feet in height and estimated to be at least 500 years old, once grew commonly in the southern part of the forest. A less attractive, somewhat scraggly species is the PITCH PINE, a small tree that can survive adverse growing conditions and thus colonizes rocky and sandy soil. It is found throughout New England and southward into the deciduous forest biome.

Subalpine Forest. After one struggles through the forests clothing the lower slopes of the western mountains, he eventually reaches a community that consists of only a few hardy veterans, a select company which has chosen to live in an environment with high winds and low temperatures. Many of the trees are twisted and dwarfed, hugging the depressions and clinging to the lee side of cliffs. This is the subalpine forest of the Rocky Mountains and the Pacific Coast Ranges. Here too the common trees are spruces, pines, and firs — but they are different species from those encountered in the boreal forest of eastern North America.

Two kinds of western pines are part of the subalpine forest: lodgepole and bristlecone. LODGEPOLE PINE is the most common member of the community, being found over a thousand-mile range from Alaska to the Rocky Mountains and the Sierra Nevada. It is a slender and graceful conifer with a narrow crown; its average height is one hundred feet or less. Lodgepole pine has a wide altitude range also, being found from sea level on the North Pacific Coast to above 11,000 feet in the southern Sierra Nevada. The flexible straight-shafted young trees were used by the Indians in making their lodges. This conifer marches in solid ranks up the mountain slopes to timberline, making its last stand amid grasses and small woody plants that edge the tundra. Among this select company of timberline veterans, the champion for longevity is the misshapen and stunted BRISTLECONE PINE. A prostrate mass of weather-

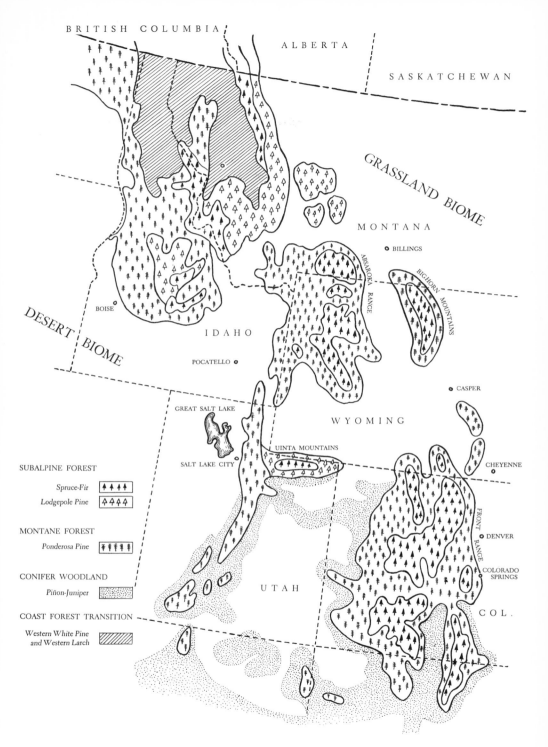

THE CONIFER BIOME: ROCKY MOUNTAIN FOREST

Subalpine conifer forest near timberline in Mount Baker National Forest, Washington. (*Credit: U.S. Forest Service*)

beaten branches, sparsely provided with needles, it reminds one of a huge piece of living driftwood. This pine grows at elevations of 8000 to 11,000 feet in the Rocky Mountains of Colorado and Utah, and is also found in isolated mountains of Nevada and eastern California. In a bristlecone pine, only twenty feet high, the spark of life has persisted longer than in any other living thing, even the redwoods. In 1953 a grove of California bristlecone pines were studied by scientists and their ages established as accurately as possible.* This revealed that seventeen of the trees were more than 4000 years old and one was a record-breaking 4600 years of age. This exceeds the oldest redwood by at least 500 years, and gives the bristlecone pine the distinction of being the

* See "Bristlecone Pine, Oldest Known Living Thing" by climatologist Edmund Schulman in the *National Geographic Magazine* for March 1958.

THE CONIFER FOREST / 141

world's oldest living thing. It is a paradox that it grows in the subalpine forest, in one of the most rigorous environments of the conifer biome.

Several spruces are members of the Rocky Mountain subalpine forest. ENGELMANN SPRUCE is a characteristic tree of the high-altitude evergreen forests, distinctive in its deep blue-green foliage and tapering pyramidal shape; typical trees are about one hundred feet in height and an estimated 500 years of age. At high elevations this spruce is associated with lodgepole pines. The undergrowth of the subalpine forest usually consists of Engelmann spruce seedlings of different ages, since the young trees are very tolerant of shade.

Mingling with the conifers are a few deciduous trees, often abundant in canyons and mountain valleys. ASPENS are very common on the lower slopes of the southern Rocky Mountains; in autumn their golden-yellow foliage adds a brilliant contrast to the dark green of the surrounding evergreens. Also growing with the spruces and pines are ARCTIC WILLOW and DWARF BIRCH, low-growing deciduous trees that often venture up to timberline. At elevations of 13,000 feet they cling closely to the rocky surface, transformed into prostrate shrubs. At these high altitudes they provide welcome forage for the herbivores occasionally found above timberline.

Montane Forest. Visitors to the national parks in the Rocky Mountains and Sierra Nevada are more familiar with this wildlife community than with the subalpine, since it is the setting for many of the camping areas and recreation grounds. This forest, being at a lower elevation than the subalpine, grows in an environment with richer soil and more favorable growing conditions. Hence the trees are larger and the forests

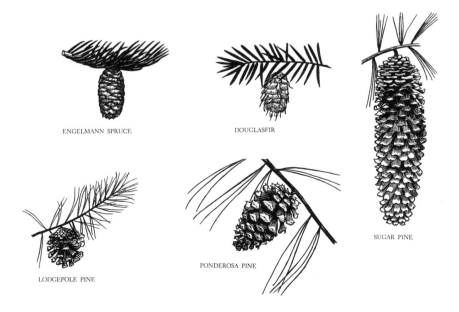

ENGELMANN SPRUCE

DOUGLASFIR

SUGAR PINE

LODGEPOLE PINE

PONDEROSA PINE

Montane forest includes pure stands of ponderosa pine. Kaibab National Forest, Arizona. (*Credit: U.S. Forest Service*)

denser. The dominant species are ponderosa and sugar pines, Douglas-fir, incense cedar, and Sierra redwood.

PONDEROSA PINE, or western yellow pine, is a stately tree with yellowish-brown trunk which rises upward fifty feet or more before the first branch appears, and terminates in a spire-like crown of dark green foliage. The thick bark separates into flat-topped plates with a loose scaly surface. Many ponderosa pines reach a height of several hundred feet, thriving on well-drained slopes and high plateaus to an elevation of 12,000 feet in the southern Rocky Mountains. Four fifths of all the trees in the forests of the Colorado Plateau are ponderosa pines. This species is so drought-resistant that it can grow where the annual rainfall is as little as fifteen inches a year; it is often the last tree to give way to the desert vegetation, at low elevations. Another tall and impressive conifer, predominantly a California tree, is the SUGAR PINE, which reaches its greatest development between 3000 and 8000 feet on the western slopes of the Sierras. Its name refers to the resin that exudes from white blisters on the furrowed bark. Large individuals reach a height of several hundred feet, with a circumference exceeding thirty feet. Its cones, often more than a foot in length, are the largest of all the conifers.

One of the most magnificent trees of the montane forest is the DOUGLASFIR, or Douglas spruce, which ascends the Sierras to an elevation of 5000 feet, the Rocky Mountains to 11,000 feet. The trunk, with deeply furrowed reddish-brown bark, forms a clean shaft for a third of its height, then terminates in a tapering crown of dark green foliage. Many Douglasfirs are more than 200 feet high; the largest individuals are found in the more humid coastal forest. INCENSE CEDAR is an at-

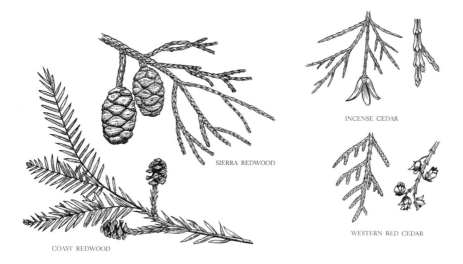

SIERRA REDWOOD

INCENSE CEDAR

WESTERN RED CEDAR

COAST REDWOOD

The General Sherman tree in Sequoia National Park is a giant Sierra redwood with a diameter of 36 feet.

tractive and aromatic tree of the California and Oregon montane forests, where it grows in groves or scattered throughout stands of ponderosa or sugar pines. Incense cedar is a tall columnar evergreen with cinnamon-red trunk and yellowish-green foliage.

The monarch of the montane forest in California is the giant SIERRA REDWOOD, or big tree. This mighty conifer grows in protected basins at elevations between 4000 and 8000 feet, surrounded by sugar and ponderosa pines and incense cedar. Today this species of Sequoia is found only in scattered groves — some seventy in all — on the western slopes of the Sierra Nevada. Fortunately most of the surviving trees will be preserved for posterity by being within the boundaries of Sequoia and Yosemite National Parks. Sequoias are a remnant of an ancient race of conifers, widespread in prehistoric times throughout North America and now making their last stand against extinction in our western mountains. Each Sierra redwood soars skyward in such graceful proportions that a viewer hardly realizes the immense girth and stature. At the summit is borne a surprisingly small crown of foliage. Well known to visitors are the Grizzly Giant tree in Yosemite National Park and the General Sherman tree in Sequoia National Park. The General Sherman tree has a diameter of thirty-six feet, a height of 272 feet. Translated into familiar dimensions, this is equal to the width of a house, and half the height of the Washington Monument. Sierra redwoods are among the oldest of all living things, some having begun their growth on the Sierra slopes 4000 years ago. Such longevity can be explained in part by their immunity to insects and fungi, and the protection afforded by a layer of spongy bark, two feet in thickness.

Coastal Forest. Unlike the subalpine and montane forests, which lie in relatively inaccessible areas, the coastal forest extends along the populated shores of the Pacific Ocean from California to Canada. If we travel the highway between San Francisco and Seattle, we find ourselves continuously surrounded by the giant trees that make up this western subdivision of the conifer biome. Here the abundant rainfall and high humidity, with moderate temperatures, have favored the growth of the most luxuriant evergreen forest of the continent. The dominant members of this community are western hemlock, western red cedar, Douglasfir, and the coast redwood.

The WESTERN HEMLOCK is a tall conifer with lustrous and feathery foliage, borne in a narrow pyramidal crown; it grows to a height of one hundred feet or more. Common to Washington and Oregon, and to a lesser extent in the northern Rocky Mountains, western hemlock

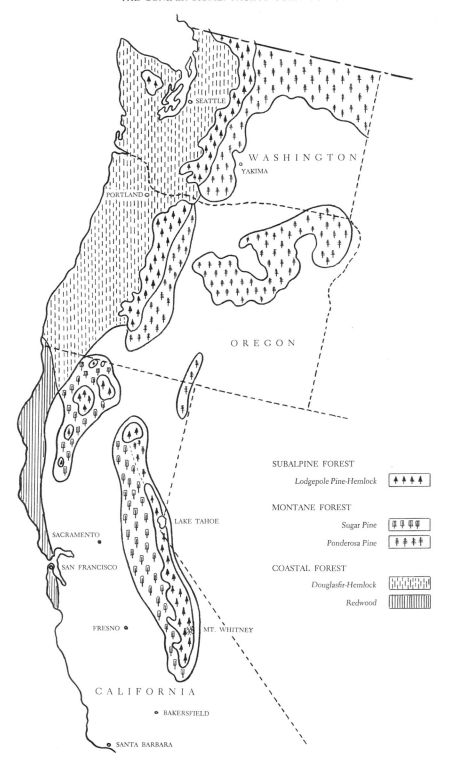

reaches its greatest development on the western slopes of the Cascade Range. Another member of the coastal forest is the WESTERN RED CEDAR, also known as giant arbor vitae, whose fluted trunk rises from a swollen base, bearing dark green foliage in flattened sprays. Giant individuals in Olympic National Park reach a circumference of sixty-two feet, a height of 150 feet. This cedar grows in moist habitats from California to Alaska, and is found up to elevations of 7000 feet in the northern Rocky Mountains.

The Olympic National Forest possesses a fine example of this coastal forest, with many giant trees of various species. Among them is a Douglasfir 221 feet in height. In the dim light of the forest floor, thriving in the heavy rainfall and mild climate, every fallen tree and boulder is carpeted with a mantle of mosses, ferns, and shade-tolerant flowering plants.

The abundant rainfall of the coastal forest provides an ideal environment for another species of Sequoia, the COAST REDWOOD. This, a taller and more slender tree than the Sierra redwood, has a shorter life-span. Its home is on a coastal strip thirty-five miles in width and 500 miles in length, reaching from northern California into southern Oregon. Coast redwoods hold the record for being the tallest of all American trees. The champion, known as the Founder's tree, growing in the Bull Creek Flat grove of California's Humboldt Redwoods State Park, has a height of 364 feet. Few thrills can equal that of seeing, for the first time, a grove of these impressive trees. There is a feeling of graceful massiveness, of airy charm combined with incredible stature as the straight trunks soar upward like pillars in a cathedral. Only occasional shafts of sunlight penetrate to the floor of the redwood groves, ruddy with a thick carpet of needles. No exploration of American wildlife communities is complete without a meeting with the coast redwood, and its high-altitude relative the Sierra redwood. They represent the ultimate in the green plant pattern of life.

The Living Environment: The Herbivores

The conifer biome, whose tree members we have just met, is the home of many kinds of herbivores. The dominant trees provide shelter and food; the foliage and fruits of the many smaller herbaceous plants that clothe the forest floor provide additional nourishment for the energy consumers which link the green plants with the carnivores. In the herbivore niches of the biome, three groups of animals play a major role: the insects, the birds, and the mammals.

Plant-feeding insects abound in the forest. Few trees are spared attack by these small but often voracious herbivores. Some insects prefer buds and foliage; such are the larvae of the many different kinds of moths and sawflies. Budworms infest the spruces of the boreal forest, tussock moths defoliate a variety of evergreen and deciduous trees. Other insects bore through the bark to feed on the underlying living tissue of the trees. These include the bark beetles found on Engelmann spruce and ponderosa pine in the West, red spruce in the East. Epidemic outbreaks of tree-infesting insects have destroyed many thousands of acres of conifer forest. Fortunately the food chain in the biome usually includes an army of insectivorous birds that keep the insect population within bounds.

A striking example of the balance maintained in a community between the insect herbivores and the insectivorous birds of the carnivore niche took place in the White River National Forest of Colorado, a forest dominated by Engelmann spruce. The animal life in the herbivore niche includes bark beetles, which feed on the inner tissues of the trunk of this tree. The beetles in turn are eaten by the arctic three-toed woodpeckers, which feed on them during the winter. Under normal conditions the beetle population is held in check by the woodpeckers and thus cannot become great enough to damage the spruces seriously. But one year a storm devastated this area, felling many of the trees. Some of the fallen trunks retained enough roothold to continue living though overthrown. Feeding all winter on the underside of the trees, the beetles were protected by the mat of crushed branches from the woodpeckers. As a result the beetle population increased, and soon hordes of insects spread to neighboring undamaged trees. The woodpecker population, ordinarily ample enough to control the beetles, could not cope with the unexpected population increase. Within a few years dead Engelmann spruces, killed by the beetles, covered thousands of acres. Economically the damage could be counted in the loss of four billion board feet of lumber. As it affected the forest community, food

RUFFED GROUSE

MOUNTAIN QUAIL

CHICKADEE

and shelter disappeared for all the smaller plants and animals. The cause of this upset was a single insignificant member of the community which had got out of control.

Only a few birds are strict herbivores, relying chiefly on plant food. These are members of the grouse and quail families. Widespread throughout the forest is the RUFFED GROUSE, a chunky and brownish bird that roosts in the conifers and feeds in the bushy undergrowth. Grouse eat seeds, beechnuts, and acorns; when the ground is covered with snow they eat buds and twigs of aspen and birch. Cutover woodlands and abandoned farms are a favorite habitat, and therefore ruffed grouse have increased in numbers since the destruction of the original eastern forests. Sprightly MOUNTAIN QUAIL, sporting a long head plume, are found in the coastal evergreen forest.

Many of the birds are omnivores, feeding on foliage, seeds, or fruits at one time of the year, adding insects to their diet in spring and summer. In this way the omnivores are also important links in the flow of energy from the green plant producers to the higher consumer levels. Some of these birds are as characteristic of the community as the spruces and firs. Memories of a summer spent in the conifer forest inevitably recall the haunting song of the white-throated sparrows, the cheerful chatter of the chickadees, the echoing notes of thrushes, and the noisy exuberance of jays. These memories also call to mind visions of colorful grosbeaks, crossbills, finches, and warblers.

Some of the forest-dwelling birds possess bills especially adapted for a diet of seeds and nuts. Such are the crossbills and grosbeaks, readily recognized by their unusual bills, which enable them to extract seeds from cones as well as to crack them open. The dimly lit aisles of the spruce and pine forests are the habitats of two northern thrushes, well camouflaged to resemble the needle-covered forest floor. The HERMIT THRUSH, found from Alaska and the Pacific Northwest to New England, is often seen on the ground looking for seeds or insects hidden in the leafy debris. SWAINSON'S THRUSH — or the olive-backed thrush as it used

JUNCO

CANADA JAY

STELLER'S JAY

to be called — has similar feeding habits but is often seen in the trees as well as on the ground. It, too, is a resident of the forests, from Alaska to the Appalachian Highlands.

Jays are a lively addition to the conifer forest community. The gray CANADA JAY, resembling an oversized chickadee, lives throughout the boreal forest; in the West it is replaced by the ROCKY MOUNTAIN JAY. Jays devour seeds, fruits, insects, and especially food left on campers' tables. Their fearless snatching of such items has earned them the name of "camp robbers." In the pine forests of both the Rocky Mountains and the Pacific Coast Ranges lives the larger, dark blue STELLER'S JAY. This is the only blue bird with a crest found west of the Rocky Mountains.

Many of the mammals are strictly plant-feeders, and consequently are significant in the food chains leading from green plants to the carnivores. In this herbivore niche we find three groups of mammals, each with teeth specially suited for a herbivorous diet. These are the rodents, the hares, and the ungulates. Many of the rodents are small members of the community which, like the mice and the lemmings, hide from view in their underground homes. Some of the larger rodents, however, are more conspicuous and frequently encountered in the conifer forests. These include a variety of tree squirrels, ground squirrels, western chipmunks, and the porcupine.

The noisy and inquisitive RED SQUIRREL is a ubiquitous tree squirrel also known as "chickaree" and "boomer." This reddish-brown forest sprite lives over a wide range from Alaska, Canada, and the western mountains to the Appalachian Highlands. It is an accomplished climber, thoroughly at home in an arboreal habitat. Red squirrels spend most of their summer days in snipping off cones of spruces; they are the biome's champion hoarders, often collecting a pile of cones 20 feet in diameter, far more than they can eat. The unused cones provide seeds for germination of new conifers; thus the red squirrel plays a role in the dispersal of the trees on which it feeds. Red squirrels also eat buds, flowers, and even mushrooms, stringing the fungi on the branches to dry before storing them. The red squirrel does not hibernate but spends long periods in sleep, alternating with short forays to its hidden food caches.

Two western cousins of the red squirrel are the sociable DOUGLAS SQUIRREL, or Sierra chickaree, of the Sierra Nevada and Cascade montane forests, and the TASSEL-EARED SQUIRREL, common in Grand Canyon

RED SQUIRREL

GOLDEN-MANTLED GROUND SQUIRREL

TASSEL-EARED SQUIRREL

National Park. The Douglas squirrel has rusty underparts rather than the whitish underparts of the red squirrel, and a bushy tail fringed with white. The larger and heavier tassel-eared squirrel has a gray or chestnut-brown coat, a fluffy gray and white tail, and large pointed ears, tipped with conspicuous tufts of hair in winter.

Chipmunks are ground-dwelling members of the squirrel tribe. Their diet includes nuts, fruits, berries, and any succulent portion of a plant; they eat seeds of all kinds, some as small as those of thistles and dandelions. Typical of the western chipmunks are the LEAST CHIPMUNK, which ranges in the montane forest eastward to the Great Lakes, and the YELLOW PINE CHIPMUNK, which lives in the ponderosa pine forests of the western mountains.

Ground squirrels are well named, for they spend most of their lives in their underground homes either sleeping or hibernating. When aboveground, a ground squirrel usually sits stiffly erect, a habit responsible for its other common name of "picket pin." Ground squirrels are familiar rodents of the western United States, adapted to a wide range of living conditions from tundra and conifer forest to arid grasslands. In the conifer biome a common representative is the GOLDEN-MANTLED GROUND SQUIRREL. It resembles a large chipmunk in its striped body but differs in having no stripes on its face. This squirrel has a wide range throughout the western montane and coastal forests. Its buff-colored body has a tawny "mantle" on the head and shoulders. Often seen around rock piles at high elevations, this squirrel enjoys basking in the sun, sitting contentedly on its haunches and resting its folded paws across its belly.

Marmots are large, heavily built western woodchucks. They enjoy startling the hiker in the western montane forest with their piercing whistles, which set the mountainsides echoing, a signal for all other members of the community to watch out for danger. The YELLOW-

BELLIED MARMOT has its burrows on the boulder-strewn slopes of the pine and spruce forests; it is a golden-brown animal with a sooty face and yellow underparts. Some individuals weigh as much as fifteen pounds. Most of its active hours are devoted to gorging on grasses and other succulent vegetation. On sunny days it takes long siestas on the warm rocks, but needs to remain alert for sudden attacks by eagles and its many furred enemies, all of which consider the fat and plump marmot a most desirable meal.

In exploring the conifer forests, one frequently meets an awkward animal with humped back, black fur, short bowlegs, and small head; its body and tail are armed with yellowish white quills which ordinarily lie concealed in the fur but which can be quickly erected to form a bristly armor. This is the lumbering PORCUPINE, which relies upon the effectiveness of its armor to discourage its enemies. With hindquarters turned to the aggressor, it raises the barbed quills and swings the spiny tail upward. Wise predators leave this strange animal severely alone; only a few, like the fisher, have learned to turn it over and attack it through its unprotected belly. The porcupine has a wide transcontinental range, inhabiting deciduous as well as evergreen forests from New England to the western mountains and Alaska. It is a large rodent, sometimes reaching a weight of twenty-five pounds. Strangely, this clumsy animal climbs trees readily and moves from branch to branch with the skill of an acrobat. In summer porcupines feed on succulent herbaceous plants and berries; in winter they chip bark from trees and feed on the inner cambium, thus damaging many forest trees.

Another familiar resident of the conifer forest is the VARYING HARE, also known as the snowshoe rabbit. This herbivore has a transcontinental range from western Canada and the Rocky Mountains to northern New England. It is a large member of the rabbit family, with longer ears and hind limbs than the cottontail's. In summer it eats grass, foliage, and buds; in winter it resorts to conifer foliage and bark of aspen or birch. The varying hare is the fleetest animal of the forest, well adapted for winter travel by large snowshoe-like feet. The toes spread wide and the soles are covered with hair. This enables it to race over snow and ice with incredible speed. Like many tundra mammals, the varying hare changes its brown summer coat to one of white in winter. The population of hares fluctuates periodically and causes corresponding changes in the numbers of predators relying upon the hare for food. Such predators are the great horned and gray owls, the gyrfalcon, and all the larger mammal carnivores of the biome.

THE CONIFER FOREST / 153

The largest herbivores of the conifer forest are the hoofed mammals, or ungulates. One of the thrilling experiences in hiking through a northern forest is to meet a moose or a caribou, or when camping in one of our national parks to encounter a deer or an elk. Deer are perhaps the most familiar and approachable large mammals of our forests, evergreen or deciduous. The common WHITE-TAILED DEER is found from British Columbia and the Rocky Mountains east to the Appalachian Highlands and south to Florida. In the northern part of its range a white-tailed deer may weigh as much as 400 pounds; southward they become smaller in size. Deer feed upon a variety of plant food in summer; in winter they have to subsist on conifer foliage, especially arbor vitae and balsam fir. They prefer open and brushy areas, usually avoiding dense forests. By destroying the original eastern forests, man has increased available deer habitats with resulting increase in the deer population, now estimated at more than 6,000,000 animals. In the western

An exciting sight, possible in some of the western national parks, is that of a migrating herd of stately elk. (*Credit: New York Zoological Society*)

United States this species is replaced by the MULE DEER, distinguished by its large ears. Mule deer are the common animal visitors at campgrounds in the western national parks, soon learning to accept food from these new members of the montane community. Mule deer are abundant from Minnesota west to British Columbia, southward through all the mountainous areas. This species too is increasing in numbers, with an estimated population of over 3,000,000.

An exciting sight, possible in some of the western national parks, is that of a herd of stately elk, a forest mammal of the northern Rocky Mountains and Pacific Coast Ranges. ELK, or wapiti, is a large deer with maned neck and huge antlers; a full-grown buck stands five feet high at the shoulders and may weigh as much as 1000 pounds. Elk herds congregate on the high slopes of the mountains in summer, migrating to lower elevations in the montane forest in winter. They feed mainly on grass but also eat foliage of shrubs and trees. In winter they browse on Douglasfir and juniper, or on twigs of aspen. Elk formerly ranged throughout the northern half of our continent, from southern New England to the Pacific Coast. Excessive hunting exterminated the great eastern herds; the last-known elk in Pennsylvania was shot in 1867. Elk today are protected in several national parks; best known is the elk refuge in Grand Teton National Park.

Less frequently encountered are the woodland caribou and moose. WOODLAND CARIBOU is larger and darker than the barren ground caribou of the tundra. This species inhabits the southern part of the boreal forest in Canada. Formerly it was also found in northern New England and the Great Lakes states, but like the elk it was exterminated many decades ago by overhunting. Woodland caribou have been reintroduced in Minnesota, and more recently in Maine. In 1963 a herd of seventeen woodland caribou were brought from Newfoundland to Maine, and air-lifted by helicopter to the summit of Mount Katahdin. Here, far from highways and interference by man, these stately mammals may repopulate an area in which their great grandparents once lived. Caribou feed on foliage, berries, and other plant food in summer; they thrive on mosses, lichens, and willow twigs in winter.

The conifer biome of North America is fortunate in counting among its inhabitants the largest antlered animal in the world. The picturesque MOOSE stands almost seven feet high at the shoulders and reaches a weight of almost a ton. Its stiltlike legs are four feet long; a moose standing on its hind legs can browse on foliage twelve feet above the ground! Yet in spite of these stalwart proportions, if all our mammals

MOOSE

were to enter a beauty contest, the moose would win no prize. It is an ungainly beast with drooping muzzle, rubbery lips, and a humped back. A pendent "bell" of hair and skin hangs at its throat. To offset these features, however, nature has rewarded the moose with a crowning glory of huge shovel-like antlers that may have a spread of seven feet. Moose rarely stray far from willow thickets, although in winter they are usually found in the shelter of conifers growing in swampy parts of the forest. They often wade shoulder-deep in the water to feed on aquatic plants and to escape the plague of flies and mosquitoes that suck their blood in summer. The favorite food of moose, in addition to willow, is foliage and twigs of maple, mountain ash, birch, and ash. In winter they have to rely upon foliage of cedar, yew, and balsam fir. Moose are found from Alaska to Nova Scotia, and at times in northern New England and the Northwest. Their number has increased with the abandonment of farms and the remains of lumbered areas, since some of their favorite food plants appear during the succession which takes place.

The Living Environment: The Carnivores

All of the animals we have so far encountered in the conifer biome are near the base of the pyramid of life. Whether bark-feeding insect, seed-eating bird, or foliage-consuming mammal, they channel the flow of energy from the green plants to the animal life of the forest. Near the apex of the pyramid of life are the animals that rely upon these herbivores for their sustenance. Some of these carnivores are insects. Ground beetles feed on caterpillars; ladybugs on aphids; aquatic bugs and diving beetles on tadpoles, small fishes, and salamanders, as well as other insects; dragonflies on small flying insects. Perhaps more evident to anyone who has hunted or fished in the conifer forest are the many bloodsucking insects. They are the mosquitoes and flies — deerflies, greenheaded flies, horseflies — which swarm through the woods in early summer. These tiny carnivores make life miserable for caribou, moose, deer, and other mammals of the boreal forest, including that other mammal man himself. All these insects, and their herbivorous relatives, however, are an important part of the diet of the many insectivorous birds.

Other consumer niches in the conifer biome are occupied by vertebrates. The cold-blooded amphibians and reptiles are primarily carnivores but, as we have seen, they are poorly adapted for life in a cold climate. Therefore they are relatively rare in the evergreen forests. Those that do occur are not restricted to this biome, but occur also in the milder deciduous forest biome to the south. Hence the comparatively few frogs, salamanders, turtles, and snakes that have emigrated northward will be described with their southern relatives in the deciduous forest biome. The most common carnivores in the conifer biome are the birds and mammals — warm-blooded animals adapted to remain active winter as well as summer.

The feathered carnivores are chiefly the woodpeckers, the hawks and eagles, and the owls. Each has its own niche and thus rarely has any conflict with other birds having similar appetites. Woodpeckers are specialists in feeding on insects that infest the bark of trees. They have many adaptations for an arboreal existence associated with their feeding habits. Clawed toes and stiff tails serve as support while they perch on or climb tree trunks. Strong sharp bills can drill through bark and wood to uncover hiding insects. And powerful head and neck muscles give the bill the impact of a small pile driver.

PILEATED WOODPECKER

BALD EAGLE

GOLDEN EAGLE

GOSHAWK

OSPREY

Those large birds of prey, the hawks and the eagles, are the top carnivores in many parts of the biome. A number of species range widely into the deciduous forest and even into the grassland biome. A common member of the community is the slaty-gray GOSHAWK, found throughout the conifer forests; its prey consists of lemmings, mice, rabbits, chipmunks, and squirrels. The OSPREY, or fish hawk, is a dark brown bird with light-colored head and underparts; it is a huge bird of prey, with a wingspread of five feet or more. Ospreys feed almost entirely on fish and so are seldom found far from lakes and the ocean. They are especially abundant along the eastern coast of North America, building huge nests of sticks and debris in the tops of conifers or on a rocky pinnacle. The BALD EAGLE is a larger bird still, with dark brown plumage and a white head. Since fish are also its preferred diet, it is usually found near the water. Bald eagles were formerly common throughout the eastern forests, breeding over a wide range from the boreal to the Louisianian life zones. But in recent years their numbers have been declining rapidly, for reasons not yet completely understood. The GOLDEN EAGLE is a bird of mountainous terrain, most abundant in western North America, where it occurs from the tundra to the southern Rocky Mountains. It is a dark brown bird of about the same size as a bald eagle, often with a wingspread of more than six feet. From its aerie on a cliff in the montane forest, it soars on air currents in a constant search for rabbits, squirrels, foxes, and even young deer. Golden eagles require fifty or sixty square miles of territory per pair, a limitation that results in these predators never being numerous in any one area.

Owls are the nocturnal hunters of the biome. Smallest is the BOREAL OWL, the size of a screech owl but lacking eartufts. Like the other owls, it preys chiefly on rodents and similar small mammals. It is a species of the Far North, ranging to treeline in the boreal forest. The GREAT HORNED OWL is a member of several biomes, with a wide range from the tundra to the deciduous forests. It is a large bird, almost two feet in length, capable of capturing such prey as mink, skunk, and grouse. The great horned owl is a barred brown bird with the most conspicuous eartufts of the entire owl family. The GREAT GRAY OWL is even larger than the great horned owl, but is a uniform gray, with a massive head. It is a circumpolar species, sometimes found as far south as the Rocky Mountains and the Pacific Coast Ranges.

Mammal carnivores add excitement to the wilderness. It is thrilling to meet a deer or a porcupine on a woodland trail, or to have a red squirrel eat out of your hand. But to come unexpectedly face-to-face with a mountain lion or a bear is an experience one is likely to remember for a lifetime. A meeting with one of these predators of the wildlife community is, however, very unlikely. Many of them are nocturnal and as a result are infrequently seen by day. Most mammal carnivores are distrustful of man, avoiding his presence whenever possible. For these reasons, the large predators are elusive animals, except where they have learned to live in close proximity to man, as in some of the national parks. While many herbivores of the biome are increasing in numbers, the carnivores on the contrary are decreasing, and retreating deeper and deeper into the inaccessible parts of the forest as civilization closes in on them. Although many of the mammal carnivores occur in several biomes, some members of the weasel, cat, and bear families are typical of the conifer forests.

Weasels are short-legged, supple-bodied carnivores ranging in size from that of a chipmunk to a small bear. Among the smaller members of the family is the SHORT-TAILED WEASEL; it weighs less than six ounces, but ounce-for-ounce this tiny carnivore is the most ferocious of all predators, relentless in its pursuit of rodents, rabbits, and birds. Like some of the tundra mammals, it changes coat with the seasons. Its fur is yellowish brown in summer but snowy white in winter, blending into the snowy background of its haunts. In its white coat this weasel is known as ermine. It is an animal of the boreal forest, being found as far south as New England and the Rocky Mountains.

One member of the family has lustrous dark brown fur so highly prized that more than a million individuals are trapped annually and five million more are raised on farms for their valuable pelt. This is the

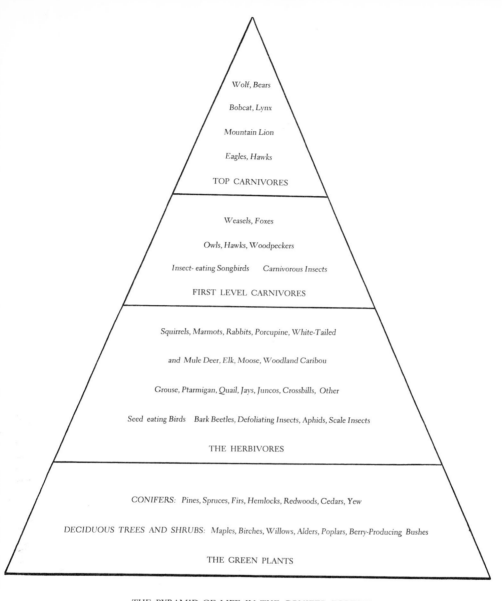

THE PYRAMID OF LIFE IN THE CONIFER FORESTS

MINK, considerably larger than the short-tailed weasel and reaching a weight of about three pounds. Mink retain their dark coat throughout the year. We can expect to find mink along the edge of streams and ponds throughout the boreal forest; they also are common over much of the rest of the country, except in the arid Southwest. Mink are expert divers and swimmers, feeding on crayfish, fish, and amphibians. They are especially fond of muskrats, which are a staple food. In the food chain of the biome they are preyed upon by larger carnivores like the bobcat and lynx.

A tree-dwelling member of the weasel family is the MARTEN, a foxlike animal with small head, broad rounded ears, and a bushy tail. The yellowish-brown fur remains the same color, winter and summer. A marten is about the same size as a mink. It is a close relative of the European sable, the American species sometimes being known also as "sable." The marten preys on red squirrels, chipmunks, rabbits, and mice; the marten, in turn, is eaten by lynx, fisher, and even eagles. Martens are inhabitants of the boreal forest and also of the subalpine-montane forests of the western mountains.

The conifer biome is the home of the two largest members of the weasel family, the fisher and the wolverine. A FISHER has silky dark brown or black fur, and may weigh as much as twelve pounds. It is more likely to be found on the ground than in the trees; bounding along with four-foot leaps, it can easily capture squirrels, marmots, and especially porcupines — its favorite prey. The fisher is a powerful fighter, often victorious in an encounter with a bobcat or a coyote. Originally fishers were found throughout the boreal forest and as far south as North Carolina, but now the only fishers in the United States occur in northern New England and New York State. Giant of the weasel tribe is the WOLVERINE, sometimes aptly called the "skunk-bear," since it can give off the offensive musk odor characteristic of all weasels and has the build of a small black bear. A wolverine is a squat animal with an arched back, bushy tail, and big feet; it grows to a length of four feet and may weigh as much as sixty pounds. An animal of such bulk and strength can prey on large herbivores like elk and moose. Today the wolverine is rare outside Alaska and Canada; a small number survive in some of our western national parks and game refuges. In Michigan, which is called the wolverine state, the last wolverine was killed more than a century ago.

Two members of the cat family prowl the shadows of the evergreen forests: Canada lynx and mountain lion. A CANADA LYNX is a huge gray and brown bob-tailed cat, reaching a length of three feet and weighing as much as thirty pounds. It has erect tufted ears, a neck ruff of long fur, ungainly legs, and large feet. Although able to climb and frequently seen in the branches of a tall tree, it usually hunts on the ground. Its yellow eyes are adapted for night vision, appearing as mere slits by day. The favorite prey of a lynx is a snowshoe hare, but included in its diet are mice, squirrels, and ptarmigan. The lynx is well adapted

The wolverine is the giant of the weasel tribe; a small number of this forest carnivore survive in western national parks and game refuges. (*Credit: New York Zoological Society*)

by its broad hairy feet to travel over snow and ice. Canada lynx is likely to be seen in any part of the boreal forest in Canada and Alaska, as well as in the western mountains. It formerly was found as far south in the eastern United States as Pennsylvania but today occurs in only a few remote areas of the northeastern United States. The MOUNTAIN LION (panther, puma, cougar, catamount — it goes by many different names) is the largest and most powerful member of the cat family in the ever-

Canada lynx is a huge bob-tailed cat with ungainly legs and large feet, adapted for travel over the snow in the conifer forests. (*Credit: New York Zoological Society*)

green forests. A full-grown individual reaches a length of seven feet and weighs as much as 200 pounds. The lithe muscular body is covered with thick reddish-brown or grayish fur. This predator has a very wide range in a variety of habitats from Florida swamps and desert plateaus to timberline in the Canadian and Alaskan wilderness. Its favorite prey is deer, but it will feed on any large mammal. Such a powerful predator has only one enemy — man. A few mountain lions may still remain in the northeastern United States, but if so they are very rare indeed.

Bears are the largest of all the forest carnivores; they are content with surprisingly small prey, considering their size. They eat ground squirrels, mice, insects, and fish. Bears are also herbivores at some seasons, feeding on grass, foliage, berries, and fruits. They have a well-known preference for sweets, and will gorge themselves on honey when they find a bee tree. The only common bear in North America today is the BLACK BEAR, the smallest member of the family. It varies in color from black to brown or light buff; the reddish-brown phase is common in the western United States. A black bear is three feet high at the shoulders and has a maximum weight of 500 pounds. Black bears are found throughout the conifer and the deciduous forest biomes, being absent from only four states (Rhode Island, New Jersey, Delaware, and Kansas). The black bear is very crafty and adaptable, qualities contributing to its presence in most of our national parks. Superficially it seems harmless, but visitors to the parks have to be constantly warned to accept the

Black bear cubs are frequently seen in Sequoia National Park, California. (*Credit: National Park Service*)

Grizzly bear. (*Credit: New York Zoological Society*)

friendly advances of this powerful mammal with caution. By gorging on food in late summer, the bear lays up a store of body fat on which it can subsist during its long winter's sleep, which lasts from October to May. The black bear is not a true hibernator and simply becomes dormant during cold weather; it then retires to its den, which often is a cave or merely shelter beneath a fallen tree.

The GRIZZLY BEAR is a much larger and more dangerous animal and also has a much more restricted range. These huge carnivores, formerly ranging throughout the montane and subalpine forests, have been hunted to such an extent that today they have had to retreat to the timberline forests or the vast expanse of the tundra, where they can lead unmolested lives. A few hundred remain in Yellowstone and Glacier National Parks, but they are still abundant in Alaska. A grizzly measures three feet high at the shoulders and weighs up to 800 pounds. It differs from the black bear in having a dished-in profile, longer claws, and light-tipped hairs that give a frosted appearance to the gray or brownish fur. Although a grizzly varies in color it never is black. They feed on rodents, insects, and plant food in season; they swim well and in summer feast on spawning salmon.

The largest living carnivore is the BIG BROWN BEAR, or Kodiak bear, a close relative of the grizzly which lives in southern Alaska and British Columbia. It is considered the direct descendant of the great cave bear that Stone Age hunters pursued many thousands of years ago. This giant mammal reaches a shoulder height of four and a half feet and weighs up to 1500 pounds. Even with its size, this bear usually avoids man, its only enemy. A big brown bear has yellowish to dark brown fur which is never sprinkled with gray; otherwise it closely resembles a grizzly bear. It feeds on mice, ground squirrels, and other rodents. Like the grizzly, it enjoys fishing for salmon and eats berries, grass, and fruit when they are available. This bear is still common in its restricted habitat, but such a large animal will undoubtedly become rare with the inevitable development of Alaska.

Man and the Conifer Biome

Primitive man, as a member of the biotic community, occupied the consumer niche with the omnivores, obtaining the necessities of life from the plant and animal members of the community. He hunted, fished, collected wild fruits. He also required clothing and shelter to protect him from the cold climate that prevails over much of North America. As long as the human population was subject to the same controls as the other life of the ecosystem, man's use of the wildlife resources did not upset the balance of nature. Thousands of years of occupation by Indians did little to change the balance of nature in North America.

New and more disturbing factors appeared when men with a new way of life colonized the continent. They too had to have food, shelter, and clothing. Thus they hunted and trapped, cut down trees for lumber, cleared land for farms, built towns. They brought with them new weapons with which to hunt, more effective tools with which to modify the environment to suit human purposes, modern methods of transportation which created railways and highways. They also brought with them new knowledge of prevention of disease and prolonging human life, so that their numbers increased at an unprecedented rate. This brought about unusual demands on the biotic community for their self-maintenance and for living space. In a very short time man became the sole top predator at the very apex of the pyramid of life.

Increasing human needs for food, shelter, clothing, and materials considered essential to civilized living meant greater and greater depletion of the resources at the base of the pyramid. This was not obvious to the newcomers in the American environment, who were overwhelmed by the superabundance of useful plants and animals. Forests were destroyed, prairies plowed under, mountains were mined, rivers were dammed, with no thought of either the effect on the ecosystem or the needs of future generations. Waste products of civilization polluted the air, the water, and the land. There was little concern for tomorrow on the part of the government or of the general public. The appalling story of this wasteful era in American history is revealed in Stewart Udall's *The Quiet Crisis*, which traces the intricate relation between man and the land in the days when the term "ecology" meant nothing to the average American.*

The nature of this interference with the biotic communities of our continent varies in the different biomes. In our exploration of the tundra we discovered that as yet man has had very little effect on altering that biome. The story is far different in many of the other biomes. For the moment let us see what it has meant to the conifer biome. Because of the climate and topography, only a small portion of the area covered by evergreen forests has proved suitable for farms, cities, or factories. In this respect the conifer biome has been more fortunate than either the deciduous forest or the grassland biomes. The impact of civilization upon this biome has been chiefly that resulting from the value of conifers as a source of lumber, wood products, and paper pulp. The white pine of the boreal forest, the Douglasfir and the redwood of the coastal and montane forests were either completely destroyed or recklessly exploited.

The northern portion of the boreal forest in Canada and Alaska is still wilderness and may remain so for many years to come. The rigorous climate, the inadequate means of transportation, the character of the forest trees, all are factors that may maintain the original biotic community. The southern portion of the forest, however, has suffered because of its accessibility and the value of its tree species. In New England the colonists cut down the forests to make cultivated fields and space for the growing villages and cities. Thus the original boreal

---

* *The Quiet Crisis* by Stewart L. Udall, Secretary of the Interior (New York: Holt, Rhinehart, and Winston, 1963).

forest disappeared, except for remote areas in northern Maine and the mountains of New Hampshire and Vermont. As our population spread westward, destruction of the boreal forests kept pace until it included those of the Great Lakes States. Today this region is mainly a second-growth, mixed forest in which succession has brought about an introduction of southern deciduous trees in the original conifer biome. The mammals of this forest were formerly the basis of an extensive fur trade, which reduced the native species to mere remnants of their former numbers. Hunting and killing of predators eliminated most of the other large mammals.

The subalpine and montane forests lie in the rugged terrain of the western mountains, a land also unsuitable for farming and the development of cities. It is also a region in which lumbering is carried on with difficulty. Because of the survival of large tracts of the virgin forests, the bird and mammal members of the community have not been depleted as they have elsewhere in the biome. The conifer forests are located in cool and scenic portions of the continent, a region to which millions of Americans throng in summer to escape the heat of the cities. The grandeur of the scenery, as well as the appeal of the wilderness, makes the montane forests the mecca of all outdoor-minded families. The abundance of fish and game attracts the millions of fishermen and hunters. For these reasons, large portions of this western forest have been set aside as national and state parks. More than half of all our large national parks are in this part of the conifer biome. This setting aside of the wilderness as a federal sanctuary seemed a wise step in sparing these areas from destruction by lumbering, grazing, mining, and other interests. At last we seemed to be making some attempt to restore the balance in an ecosystem which we had so carelessly upset. But even this has brought on problems, owing to conflict of ideas as to how best to make use of these protected areas. Many of the visitors wish to enjoy the public lands in the condition for which they were originally intended: as wildlife communities, unchanged from their primitive and original status. Others come to seek enjoyment in another way, thinking of the parks as vast outdoor playgrounds. They desire smooth highways, luxurious accommodations, swimming pools, and television. The paradox of these conflicting desires is that the wilderness we set aside, sparing destruction by private citizens, is in danger of being destroyed by a public uneducated in the appreciation of nature and the inexorable laws of an ecosystem.

The coastal forest is exposed to another hazard. It is a lumberman's paradise. And so this portion of the conifer biome has seen considerable destruction of the virgin forests. Fortunately, the majority of the lumber interests have developed great foresight in applying modern scientific methods to the cutting of timber, and to replanting the cutover areas to guarantee a future yield. In addition, public-spirited citizens have succeeded in rescuing many scenic forested areas, making them state and county parks. Outstanding have been the accomplishments of the Save-the-Redwoods League, with thousands of members, in California. Scattered along the famous Redwood Highway are many groves, some large and others small, acquired by various organizations that saw the need of saving these rare trees of the coastal forest community.

The conifer biome means many different things to different people. It has been in the past, and continues to be, the source of our most valuable wood products. It is the home of many unique American animals, some of which came dangerously close to extinction because of their value as food or as producers of fur. It is a wildlife community set in spectacular surroundings of mountains and gorges, lakes and rivers. Thanks to the foresight of conservation-minded individuals from Gifford Pinchot and Theodore Roosevelt to John Muir and Aldo Leopold, large tracts of this biome are today within the boundaries of many national parks and forests. In these we can see today the America that was, and become part of a biotic community that has existed for thousands of years — even if our experience is only one of a few days or a week. It is a wonderful legacy from previous generations which we must pass on unchanged and unharmed to future generations.

10 / The Deciduous Forest

Throughout the eastern United States the rainfall, temperature range, seasonal climate, and topography favor the development of deciduous trees as a climax forest. Northward along the Canadian border the shorter growing season and lower winter temperatures are more favorable to the growth of conifers; southward toward the peninsula of Florida the milder year-round temperatures and absence of a dormant season encourage the development of broad-leaved evergreen trees. A deciduous forest similar to that of the eastern United States occurs in eastern Asia, middle Europe, and South America. None, however, is as extensive or as luxuriant as the forest that originally formed an uninterrupted wilderness of some 2000 miles from north to south, more than 1200 miles from east to west. Now this vast area has become thirty-one states, and comparatively little forest remains. For in these states three quarters of the population of the entire United States have their homes. Here are our greatest industrial centers and our largest urban developments. In the face of such expansion of civilization, the extensive forests had to give way until only a remnant of the original woodland remains.

The heart of the deciduous forest biome lies in the southern Appalachian Highlands; typical of this forested area is Great Smoky Mountains National Park on the crest of the mountains forming the boundary between Tennessee and North Carolina. More than 1300 species of deciduous trees, shrubs, and other angiosperms are the members of this community. Northward the biome merges with the conifer forest, this resulting in a mixed growth of evergreen and deciduous trees. Eastward on the piedmont and coastal plain the forest becomes a mixed conifer-deciduous community that extends to the Atlantic Ocean. Southward the forest gradually merges with the tropical vegetation of the Gulf of Mexico. At its western edge, approximately on a line from Minnesota to Texas, the biome resembles a woodland park with scattered trees and prairie vegetation.

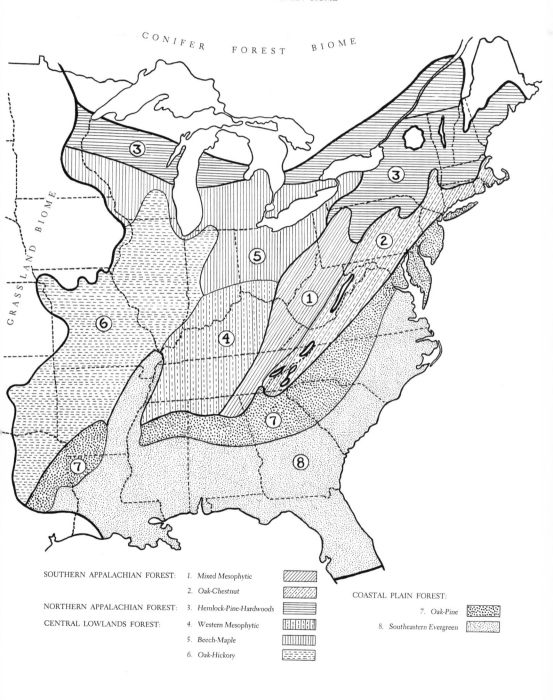

The deciduous forest is dominated by broad-leaved angiosperm trees such as these oaks and beeches.

The Physical Environment

As would be expected, so large an area presents a variety of environments. The rocky ridges, high summits, steep ravines, and sheltered coves of the Appalachian Highlands provide many different habitats for wildlife. The rushing streams make suitable homes for great numbers of hydrophytic plants and moisture-loving animals. The dense woods of the alluvial lowlands provide shelter and food for many kinds of birds and mammals. Innumerable swamps, some of vast dimensions, are ideal for hydrophytes, amphibians, reptiles, and aquatic birds.

The considerable north-south extent of the biome, from Canada to the Gulf of Mexico, offers a great range in temperature. At its northern limit, the average January temperature is 20° F.; at the southern boundary it is 55° F. Snow falls in all parts of the region, varying from a few inches annually in the south to 60 inches or more in the north. This has a marked effect on the animal life. The entire biome lies east of the 20-inch isohyet; therefore the rainfall everywhere is ample for tree growth. However, due to the great east-west extent of the forest, considerable variation above this minimum occurs. The precipitation, forty inches a year in the Mississippi River Valley, increases eastward in the mountains. The heaviest rainfall, eighty inches annually, is in the southern Appalachian Highlands. As a result, the densest and most luxuriant portion of the deciduous forest is located in this region. In the coastal-plain lowlands surplus water keeps the ground continually moist and creates a succession of swamps from Virginia to Florida.

When exploring the conifer forest, we discovered that local differences in topography, soil composition, rainfall, and temperature caused development of smaller communities — or associations — within the biome. The same is true in the deciduous forest. The result is a number of associations, each a community in itself, with characteristic species as the dominant trees. (1) A *southern Appalachian forest* forms the core of the biome, with mesophytic trees forming the climax vegetation. From this central region the forest has spread outward in all directions. (2) In the cooler portion of the biome is a *northern Appalachian forest*, also of mesophytic trees but made up of species hardy enough to establish themselves in this climatic region. (3) A *central-lowlands forest* forms the western portion of the biome, occurring in the river valleys where the soil is fertile and the topography less rugged. Beyond the Mississippi River, this western part of the forest includes

The deciduous forest reaches its greatest luxuriance on the slopes of the Great Smoky Mountains of Tennessee and North Carolina. View in Great Smoky Mountains National Park from Klingmans Dome. (*Credit: National Park Service*)

many xerophytic plants and such animals as are able to live under arid conditions. (4) A *coastal-plain forest* lying in the Louisianian life zone makes up the eastern and southern part of the biome. The vegetation is almost entirely dominated by southern conifers, and this results in a southeastern evergreen forest even though it is part of the deciduous forest biome.

The *southern Appalachian forest* reaches its greatest development, in number of species and luxuriance of growth, in the Cumberland Mountains of West Virginia, the Blue Ridge Mountains of Virginia, and the Great Smoky Mountains of Tennessee and North Carolina. Here the forest is made up chiefly of oaks, walnuts, hickories, birches, beech, chestnut, and tuliptree. A lower story of smaller trees includes flowering dogwood, redbud, and holly; beneath these may be a thicket of rhododendron, laurel, and azalea. One conifer — the eastern hemlock — is also a member of this community.

Oaks are very common, as they are in all other subdivisions of the biome, because of their great adaptability to all types of habitat from dry sandy plains to coastal swamps. They are also very resistant to disease, and so become long-lived members of the community, often reaching an age of 800 years or more. Acorns are an important part of the diet of many forest birds and mammals; thus oaks are vital food producers for the ecosystem. More than twenty common species of oak are found in the southern Appalachian forest, but most likely to be encountered while exploring this part of the biome are red oak, black oak, and white oak. RED OAK reaches an average diameter of three feet and a height of one hundred feet. The BLACK OAK is very similar in appearance and size, although more limited in range than the red oak.

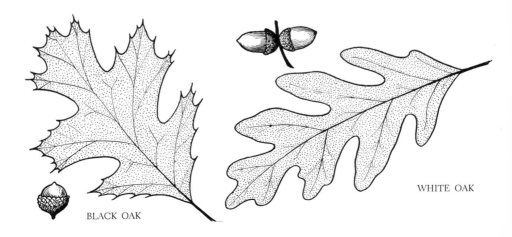

BLACK OAK

WHITE OAK

It reaches its best development in the Ohio River Valley. Since the black oak is crowded out of the better sites by other mesophytic oaks, we find it most often in poor soil, where others cannot compete. WHITE OAK flourishes on rich, well-drained soil and is abundant on the western slopes of the Appalachian Highlands. The patriarch of the white oak clan is the massive Wye Oak of Maryland's eastern shore, with a girth of twenty-seven feet.

We notice several kinds of nut trees in the southern Appalachian forest; nuts, like acorns, are a staple food item for many of the forest herbivores. BLACK WALNUT occurs on alluvial soils in the central part of the forest, being absent both in the Far North and in the South. The thick-husked fruit contains the familiar and flavorsome walnut meat. On poorer soil we find the BUTTERNUT, a close relative of the walnut. This species can endure lower temperatures and so is found in the northernmost part of the biome. Most widespread of all the hickories is the SHAGBARK HICKORY, a tall, slender tree easily recognized by the loose shredding bark, which hangs in strips on the trunk.

Another group of catkin-trees found in the southern Appalachian forest is that of the birches. As in all birches, the trunk is marked by elongated horizontal lenticels, or breathing pores. BLACK BIRCH, also called sweet birch because of the wintergreen-flavored bark, does not venture far onto the coastal plain or westward into the lowlands. YELLOW BIRCH prefers moister habitats, and is particularly common along stream margins. More northern in range than the black birch, it is found in the northern Appalachian forest.

We cannot wander far in the deciduous forest without seeing an impressive tree with smooth gray trunk and a wide-spreading crown. This is the AMERICAN BEECH, a catkin-tree related to the oaks. Its maximum

BLACK BIRCH

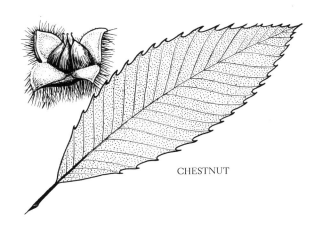
CHESTNUT

growth is attained on the western slopes of the mountains and on the alluvial lowlands of the river valleys. A favorite habitat is on the sides of shaded, moist ravines. The fruit is a small prickly bur that splits open to reveal several triangular sweet nuts, relished by squirrels, deer, raccoons, and game birds. Another catkin-tree is the AMERICAN CHESTNUT, once a common member of the Appalachian forest, ranging northward into New England. Chestnut reached its best growth on the well-drained slopes of the southern mountains but, as we know, the chestnut blight has eliminated most of the trees from the biome. Many standing bleached and dead trees are the only evidence of their former dominant role in the community.

Maples, like oaks are common everywhere in the deciduous forest biome. SUGAR MAPLE, found in every state east of the Mississippi River, thrives on rocky hillsides as well as on richer and more fertile soil. Sugar maple is also a common deciduous tree of southeastern Canada. We find the SILVER MAPLE at lower elevations, usually edging riverbanks in moist lowlands. Even more hydrophytic is the RED MAPLE, a species partial to swamps and stream margins. This maple has a very wide range; we can meet it from the swamps of Maine to the Everglades of Florida.

Outstanding among all the tree members of the southern Appalachian forest is the stately TULIPTREE, or yellow poplar. The clean straight trunk of this beautiful tree rises fifty or sixty feet into the air before the first branches appear. It grows largest in the sheltered coves and valleys of the southern mountains. A relative of magnolia, it has showy flowers several inches in diameter; the common name of the tree is suggested

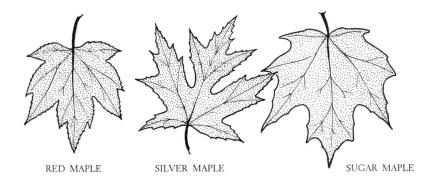

RED MAPLE   SILVER MAPLE   SUGAR MAPLE

THE DECIDUOUS FOREST / 175

by the resemblance of the bright yellow and orange blossoms to tulips. During the last geologic era, tuliptrees were widespread throughout North America and Europe, but today are represented only by this American species and another in China. The tuliptree is without doubt one of the aristocrats of the deciduous forests.

Two conifers occupy very different habitats in the southern Appalachian forest; the eastern hemlock of moist slopes and ravines, the red cedar of sandy and rocky hillsides. The EASTERN HEMLOCK has a massive trunk, reaching a diameter of six feet, and a stature of 150 feet or more; the seedlings can develop in dim light and so hemlocks spring up abundantly on the forest floor among the deciduous trees. Hemlock prefers a cool habitat and therefore is plentiful in the mountainous northern Appalachian forest. RED CEDAR is a small tree, usually only a few feet in diameter and under forty feet in height. The dark blue berry-like fruits are eaten by birds and many small mammals. Red cedar thrives on xerophytic areas where other trees cannot establish themselves; thus it plays an important role in reforesting abandoned fields and barren areas.

Beneath the tall canopy trees of the deciduous forest grow a number of smaller specimens. The showy white blooms of the FLOWERING DOGWOOD add a bright note to the eastern forests in spring. Dogwood is a small tree, rarely more than forty feet tall; it thrives on rich well-drained soil throughout this eastern forest and also in the lowlands east and west of the mountains. REDBUD is another small tree of similar size, common along streams and on fertile bottomlands. A member of the legume family, it has pealike flowers and a small pod of fruit. The

EASTERN HEMLOCK

AMERICAN HOLLY

evergreen AMERICAN HOLLY is a shrub or small tree in both the highlands and along the coastal plain; it has yellowish-green spiny leaves.

Venturing northward into the Great Lakes states, New York, and New England, we find ourselves in the *northern Appalachian forest*. This forest covers all but the summits of the White Mountains, Mount Katahdin, and the Adirondack Mountains. It completely clothes the Green Mountains of Vermont and the Catskills of New York. At higher elevations and on its northern flank in Maine and Canada, it borders the boreal forest. This creates a mixed community where white pine, red spruce, and balsam fir keep company with maples and oaks. Many of the deciduous trees we discovered in the southern Appalachian forest occur here also. As we follow a trail through the woods of New Hampshire or New York, we meet again the red oak and the white oak, shagbark hickory and yellow birch, sugar maple and red maple, beech and hemlock. But we notice the absence of some species. We find no tuliptree, chestnut, or black walnut; no flowering dogwood, redbud, or holly. On the other hand, we encounter some new species. Among these are gray and paper birch and aspen poplar.

Two species of birch are cold-climate trees and therefore common in the northern Appalachian forest. GRAY BIRCH is the smallest of the birches. It is a short-lived tree, often found colonizing cutover or burned land, where it is a pioneer in succession. PAPER BIRCH, or canoe birch, is one of the aristocrats of the northern forest. The chalky-white bark and delicate foliage are a contrast to the dark green of the spruce and pine with which it frequently grows. It grows best in sheltered moist locations, but can also survive near timberline on exposed mountain slopes.

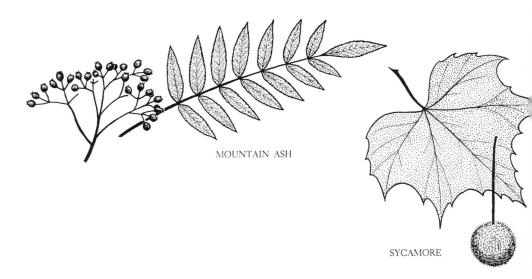

MOUNTAIN ASH

SYCAMORE

ASPEN POPLAR, like gray birch, is an important pioneer in reforesting cutover and burned areas. It springs up quickly, forming thickets on abandoned land. It is a small tree, with smooth greenish gray bark. Common in southeastern Canada and New England, aspen grows also in the western mountains, where it forms extensive communities on the lower slopes. The tender bark is the main food of beavers.

The *central-lowlands forest* occupies the westernmost portion of the biome, from the lower slopes of the Appalachian Highlands to the bottomlands of the Ohio and Mississippi River Valleys. Many of the trees of this forest are the same as those of the higher uplands, but the less rugged topography and more fertile soil combines to produce a habitat suited to species different from the predominant trees of the uplands. In Ohio, Indiana, Illinois, and Missouri we find many trees that have extended their range west from the mountains into the lowlands: sugar and red maples, walnut and hickories, beech, tuliptree, dogwood, redbud, and numerous kinds of oaks. But with them we find additional species which, although found in similar habitats throughout most of the eastern states, here reach their greatest development. Among them are sycamore and cottonwood. Several species with showy flowers are also present — buckeye, catalpa, and black locust.

A tree often seen along the banks of streams is the SYCAMORE, easily recognized by the mottled bark. Sycamore trees reach a diameter of ten to twelve feet and are so massive in girth that they are entitled to the rank of being largest of all our deciduous trees. Also following river courses and in wet lowlands is the EASTERN COTTONWOOD, a relative of the aspen poplar. The cottonwood is almost as large as the sycamore. An extensive but shallow root system anchors the tree in porous soil so effectively that it can withstand high winds; therefore it is used in shelterbelt planting.

This portion of the deciduous forest has several trees conspicuous for their large and showy flowers. One, the NORTHERN CATALPA, is found from Illinois to the southern Mississippi River region. The large heart-shaped leaves often grow to a length of twelve inches, the largest simple leaf found on any American deciduous tree. Catalpa has large clusters of white flowers that produce cigar-shaped pods filled with winged seeds. OHIO BUCKEYE is another tree with conspicuous blossoms, produced in erect clusters of yellowish-white flowers. Buckeye is related to the introduced horse chestnut and has similar compound leaves and prickly fruits, each containing a single large brown nut. This species occurs

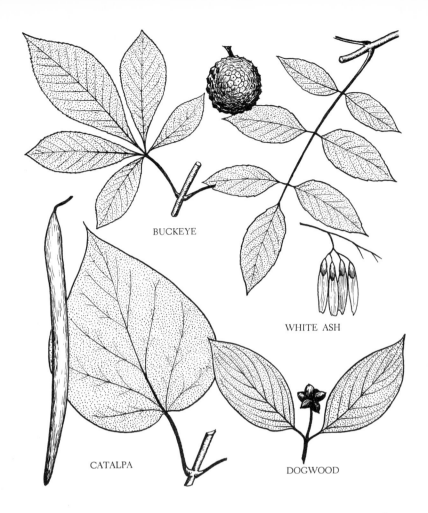

BUCKEYE

WHITE ASH

CATALPA

DOGWOOD

throughout the central states but is so abundant in Ohio that this state has long been called the Buckeye State. Originally found chiefly in the central-highlands region is the BLACK LOCUST. Its flowers form showy pendent white clusters; the branches are provided with short spines in pairs. Since it sprouts from the root, black locust often forms thickets.

As we go farther west, the forest merges into either boreal forest or grassland. In the northwestern part, surrounding the Great Lakes, the trees are chiefly beech and sugar maple, with smaller numbers of elm, hickories, oaks, and hemlock. In the extreme Northwest, sugar maple still remains a dominant tree, growing with oaks, hickories, and jack pine on the sand plains, with red pine on the rocky slopes, and with white pine on the moister hillsides. If we hollow the deciduous forest to its limits we find ourselves in the prairie of Nebraska. Here East meets West: on the low ground are the walnut and white ash of the deciduous forest, on the ridges are grooves of ponderosa pine, the outposts of the Rocky Mountain forest.

Southwestward the forest spreads into an arid area with milder temperature. This is typical of the plains of eastern Oklahoma and Texas, the uplands of the Ozark and Ouachita Mountains. Here the forest community consists of stunted oaks and hickories, with dogwood and redbud scattered among the oaks on the more mesophytic sites. On the rocky ridges and barren slopes grow thickets of red cedar. Finally, on the open plains the oaks and hickories become scarcer, and mesquite and cacti take over — an indication of the vegetation in the country to the west of the deciduous forest.

East of the Appalachian Highlands is the last, and in some ways the most distinctive, subdivision of the biome: the *coastal-plain forest*. In the preceding three associations, the dominant trees have been deciduous, with conifers (such as hemlock and red cedar) playing a minor

Palmetto, a shrubby relative of the cabbage palm, forms a dense undergrowth in the lowland pine forests.

role in the community. Between the highlands and the sea, the forest is one in which these roles are reversed. The dominant trees are conifers; the deciduous trees are in the minority. Because of the great number of conifers, this coastal portion of the biome is sometimes known as the southeastern evergreen forest. Ecologists believe that this association is a temporary climax, a consequence of many years of disturbance by fire, and that, if left undisturbed, the vegetation would reach the oak-hickory climax characteristic of the rest of the deciduous forest.

On the piedmont slopes adjacent to the southern Appalachian forest grow many of the trees we have already encountered in higher and more northern areas, mixed with some of the conifers that become more abundant on the coastal plain. This is an oak-pine community typical of western North and South Carolina; here we find our old friends beech, sugar maple, many different oaks and hickories, dogwood, redbud, and holly. In the river valleys grow cottonwood, sycamore, birches, and willows. As we approach the coast however these trees give way more and more to several kinds of southern pine, live oak, and southern magnolia. In the swampland is a dense jungle of hydrophytic trees, dominated by bald cypress. Overgrown with epiphytes such as Spanish moss, banked with palmetto and draped with vines, the forest of this part of the biome is far different from the beech-hemlock and sugar maple–tuliptree communities of the northern part of the deciduous forest. The influence on vegetation of the Louisianian life zone is evident wherever we look.

The traveler from New York to Florida drives for hundreds of miles through groves of tall pines extending in all directions. This pine forest consists chiefly of longleaf, loblolly, and slash pine. On high sandy ground grows the LONGLEAF PINE, or southern yellow pine. It is a handsome tree with clean, straight trunk, the orange-brown bark of which splits into large rectangular plates. The dark green foliage glistens in the sunlight; the needles are unusually long, sometimes eighteen inches. Longleaf pine, common from the Carolinas to Louisiana, is not only a beautiful pine but a valuable one. The trunk is tapped for turpentine and resin products, both economic assets of the South. In wet depressions thrives the LOBLOLLY PINE, a dominant tree of the Southeast; it often invades abandoned fields and thereby earns the name of oldfield pine. The tall straight trunk, covered with furrowed, cinnamon-colored bark, grows to the same height as the longleaf pine. SLASH PINE grows in a narrow belt along the Gulf coastal plain and in Florida, where, often occurring in pure stands, it is tapped for resin. The tall reddish-

The southern pines are tapped for their resinous sap, used in manufacturing turpentine and resin products.

brown trunk bears a small foliage crown of shining needles.

The drier habitats of the coastal-plain forest support two broad-leaved evergreen trees typical of the southeastern states: the eastern live oak and the southern magnolia. EASTERN LIVE OAK greets the traveler through the coastal forest both in the open country and along the city streets, from Virginia to eastern Texas. It is never found far from the seacoast, and establishes itself even on the windswept sand dunes edging the Atlantic Ocean. Live oak has a short but massive trunk frequently ten or eleven feet in diameter; the low spreading crown is often large enough to shade an acre of ground. Huge limbs extend horizontally, supporting aerial gardens of mosses, ferns, and air plants in a riotous profusion. The small leathery leaves are usually edged with sharp teeth; the acorns provide food for many forest animals. The most distinctive tree of the

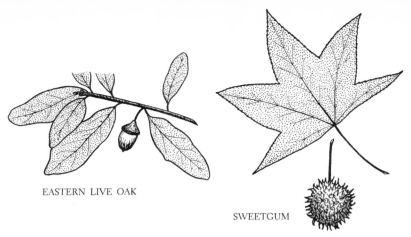

EASTERN LIVE OAK

SWEETGUM

forest is the SOUTHERN MAGNOLIA, with large glossy-green leaves and fragrant flowers, creamy white and six inches in diameter. The cone-shaped fruit is covered with scarlet seeds prized by many of the forest herbivores. This member of the community has an illustrious lineage, in preglacial times having been widespread from Greenland to Siberia.

The coastal area includes many swamps, of which the best known are the Dismal Swamp of North Carolina, Santee Swamp of South Carolina, and Okefenokee Swamp of Georgia. Swamp forests are jungles of vegetation in which live many animals requiring a humid, mild climate. Here thrives that most hydrophytic of all conifers, the BALD CYPRESS. The grayish-brown trunk has a wide flaring base that tapers to support a narrow crown of light green foliage much like the hemlock's; in winter the bald cypress is leafless. A huge cypress at Sanford, Florida, has a

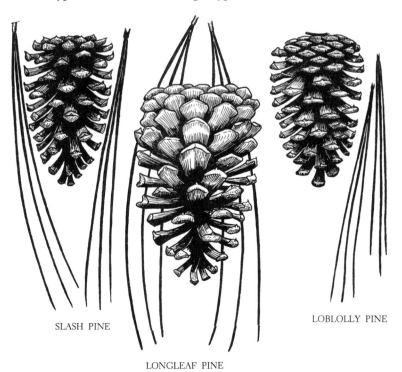

SLASH PINE

LOBLOLLY PINE

LONGLEAF PINE

The eastern live oak is often draped with Spanish moss and other epiphytes. Live-oak hammock near Ocala, Florida.

base diameter of twelve feet, is 126 feet high. In prehistoric times the range of this species was much more extensive than it is today; the few survivors of a century of lumbering are only a remnant of a race once abundant throughout North America and northern Europe. In the dimly lit, impenetrable swamp, each cypress rises from the black water, its base surrounded by a cluster of cypress knees. At the base of one of the buttressed trees we may see a sleepy cottonmouth lying ready, if disturbed, to slip into the still water; a white egret perched perhaps

In the swampland is a jungle of hydrophytic trees dominated by bald cypress, overgrown with festoons of Spanish moss. Florida.

on a moss-draped limb. The distant bellow of an alligator only adds to the feeling of a primitive wilderness.

Associated with the bald cypress is our old friend the red maple. There is also an additional deciduous tree, the SWEETGUM. Its name refers to the yellowish fragrant resin that exudes from the bark. Sweetgum has an excurrent habit of growth, more like a conifer than an angiosperm.

THE LIVING ENVIRONMENT: THE HERBIVORES

The trees of the deciduous forest, like those of the conifer biome, provide shelter for the forest animals, many of which lead an arboreal existence. Some trees in addition supply food: the catkin-trees with their acorns and nuts; the dogwood, magnolia, and wild cherry with their juicy fruits; the aspen and willow with their bark and buds. Many of the herbivores that feed directly upon this plant life are invertebrates, rarely seen but important in the flow of energy from the plant producers to the carnivores. They include millipedes, worms, and numerous ground-dwelling insects, some of which can be found under every decaying log or in each pile of forest litter. Other insects are the many aquatic species which, both as larvae and adults, are part of the food chain lead-

CARDINAL

RED-EYED VIREO

CATBIRD

MOCKINGBIRD

BLUE JAY

ing to the amphibious carnivores — the frogs, snakes, and turtles. More evident is the presence of hordes of insects that feed upon the leaves and bore into the trunks of the trees. The thin, broad leaves of angiosperms are more susceptible to insect attack than the tougher needles of conifers; thus the deciduous forest shelters many leaf-eating insects among its inhabitants. Every deciduous tree and shrub becomes a banquet table for some kind of insect. As we have already discovered, were it not for the voracious appetites of insect-eating birds, the herbivorous insects would destroy far more vegetation than they do. Such insects include the tent caterpillars, the tussock moth and gypsy moth, the gall insects, the leaf beetles, the hickory bark beetle, and the locust tree borer.

The vertebrate herbivores are mainly birds and mammals. As in other biomes, we find few strictly herbivorous birds. The majority are omnivorous, eating both plant and animal food according to which is available. Among the songbirds of the deciduous forest which live chiefly upon vegetation are the red-eyed vireo, blue jay, cardinal, mockingbird, and catbird. The largest herbivorous species are the game birds: wild turkey, bobwhite, and ruffed grouse.

Wild turkey. (*Credit: A. D. Cruickshank, from National Audubon Society*)

The largest North American game bird is the WILD TURKEY. It is a handsome bird, with bronzed-brown plumage, expansive tail, and a bare head adorned with red wattles. Wild turkeys feed on acorns, seeds, roots, and tubers, and attain a weight of thirty pounds. The Spaniards brought wild turkeys to Europe and domesticated them. The English colonists therefore had already become familiar with the turkey and, when they came to New England, were delighted to find wild turkey in abundance. To them, as to the Indians before them, the turkeys were a providential food supply. Hunting soon depleted the wild turkey population, and destruction of forestland to make farms eliminated much of the turkey habitat. Today wild turkeys are found only south of Pennsylvania; a variety lives in the foothills of the Rocky Mountains, in Colorado and Arizona.

The BOBWHITE quail is a smaller game bird, rusty brown, and with a small feathered crest. It is an elusive bird, hiding in the tall grass of brushy pastures and open woodlands. Bobwhites feed on seeds, fruits, and foliage. The increasing abundance of cornfields in the northern part of the biome has meant that there — in the discarded or overlooked corn left in the fields — winter food is available. This has encouraged the bird to extend its range farther north than the original limit. Today bobwhites can be found from Maine to Minnesota, south to the Gulf of Mexico.

Bobwhite. (Credit: Maslowski–Goodpaster, from National Audubon Society)

Three of the original bird herbivores of the biome were unable to adjust to the spread of civilization: the heath hen, passenger pigeon, and the Carolina parakeet. The HEATH HEN, an eastern variety of the prairie chicken, was a game bird of open woodlands throughout the entire deciduous forest biome. In the oak and pine communities it found ample supplies of acorns, seeds, berries, and wild fruits. The heath hen was eliminated from the eastern mainland in 1835; a small population persisted on Martha's Vineyard, an island off the coast of Massachusetts, until 1932. Then the last survivor died and the heath hen became extinct. Today the related prairie chicken is common west of Indiana, having found a more congenial home in the grain-raising prairie states.

When the colonists first explored the deciduous forest, they found an unbelievable multitude of one of the finest and largest pigeons the world has ever seen. The PASSENGER PIGEON was a pearly-gray bird with lighter pinkish breast, and a two-foot wingspan. It found ample food in the abundance of acorns, beechnuts, seeds, and fruits of the deciduous forest. At no time has this continent ever seen a greater concentration of individuals in a single species. Millions of birds could be found roosting in one wooded area. A single migratory flight, according to such reliable observers as Audubon and Wilson, included an estimated *two billion* passenger pigeons, roaring past the sun in a darkening cloud that took fourteen hours to pass! Yet by 1878 the birds were reduced from millions to thousands, and in 1914 the last passenger pigeon died in a Cincinnati zoo. What caused this sudden extinction of an apparently successful species of the community? There seem to be several possible explanations. Added to their extensive slaughter for food, the destruction of their habitat and food supplies by encroaching civilization certainly played an important role; the pigeons may have starved to

PASSENGER PIGEON

CAROLINA PARAKEET

death. Other explanations are that they contracted a fatal disease from the introduced domestic pigeons and that their reproductive rate was low — each female passenger pigeon laid but a single egg. Probably no one cause alone accounts for their extinction, but the combination of all these factors was too much even for such a populous race.

The deciduous forest biome was once the home of a brilliantly colored parrot, the CAROLINA PARAKEET. This large bird had green and yellow plumage, a long tail, and orange-tinted head. Its home was in river bottom forests from New York to Texas, where it lived on beechnuts, pecans, fruits, and seeds. Like the passenger pigeon, it traveled in flocks — making its destruction more certain when hunted for food, for plumage to decorate women's hats, or for capture to sell as a cage bird. As a result, very few survived into this century. The last Carolina parakeet was seen in the Everglades of Florida in the 1920's. It is possible that a few may survive in some inaccessible swamp of the coastal forest, but the odds are greatly against it.

The white-tailed deer is a common mammal herbivore of the deciduous forest, as well as the neighboring boreal forest biome. (*Credit: New York Zoological Society*)

The mammal herbivores include some species found also in the conifer biome. A few are the white-tailed deer, porcupines, and red squirrels. In addition we find some species more typical of the deciduous forest biome. Among them are the cottontail rabbit, muskrat, eastern chipmunk, gray and fox squirrels, woodchuck, and beaver.

Few large herbivores remain in the deciduous forest today. Most familiar are those which have adapted to the changes wrought in their habitat by human activities. Among these none is more frequently encountered than the COTTONTAIL rabbit, which we often come upon unexpectedly in a grassy glade or weedy thicket, or may see bounding across a field, its powder-puff tail flashing a white signal of danger. Cottontails are widespread throughout the eastern United States, but northward they decrease in numbers and their niche is taken over by the varying hare of the boreal forest. The brown or grayish fur of a cottontail remains the same color throughout the year. They eat all kinds of succulent vegetation, but prefer grass, clover, alfalfa, and tender garden vegetables. Their prolific reproductive rate keeps the rabbit population at a constant high level, in spite of the fact that rabbits provide more sport for hunters than all other game species combined.

Largest of the rat tribe is the MUSKRAT, found throughout the biome wherever water and an abundance of hydrophytes for food are. The diet of a muskrat consists chiefly of cattails, rushes, and other succulent aquatic plants. A muskrat looks more like a beaver than a rat — except for its tail. The glossy brown fur has a softer underfur, which provides the muskrat with a waterproof and comfortable coat for winter. Because the fur is so valuable, many millions of muskrats are trapped annually. Muskrats are active at all seasons, swimming beneath the ice the winter long. They construct large cone-shaped homes of sticks and bits of vegetation in water shallow enough to permit the top to project above the surface of the pond.

A walk through any part of the deciduous forest will usually bring us face-to-face with several members of the squirrel tribe. The EASTERN CHIPMUNK, although often seen in trees, is actually a ground-dwelling squirrel. It is a pert little animal, attired in a reddish-brown coat striped neatly with black and white on the face and sides of the body. We find this eastern species throughout the deciduous forest, and northward into Canada, where it associates with the animals of the boreal forest. Chipmunks are an intriguing combination of curiosity and shyness, one moment accepting our advances, the next dashing to the safety of a nearby rock pile. They spend the night in their dens, and forage for nuts and berries by day. Chipmunks take naps of several weeks' duration in winter, waking occasionally for a meal from the cache of food stored close to their underground sleeping quarters.

The GRAY SQUIRREL, a tree squirrel larger than the chipmunk, feeds on acorns, nuts, seeds, fruits, and mushrooms. More friendly than the chipmunk, the gray squirrel is the chief attraction at all campgrounds in the eastern national parks. Although found in every state east of the Mississippi River, it is most abundant in the Carolinian and Louisianian life zones. This squirrel indulges in long naps during the winter, interspersed with waking periods for raiding its food caches. It dens in tree cavities and, in the milder parts of the biome, in leafy tree nests.

Unafraid of man, the eastern WOODCHUCK is one of the few members

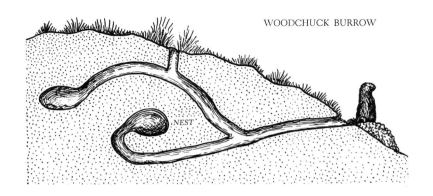

WOODCHUCK BURROW

of the wildlife community which can be seen along the highways, usually sitting erect on a grassy bank near the opening of its burrow, watching the traffic go by. A woodchuck is a large chunky rodent with brownish or grizzled gray fur; some old adults are longer than twenty-four inches and may weigh as much as ten pounds. Woodchucks are found throughout the biome except in the coastal plain section; they also range northward throughout the boreal forest. They prefer to live in sandy, well-drained habitats or open woodlands, near a source of grass, clover, alfalfa or other juicy foliage. In winter they become true hibernators.

The largest North American member of the gnawing clan is the BEAVER, in many ways an unusual member of the wildlife community. Beavers are not restricted to this biome alone, being found throughout North America wherever suitable waterways exist for construction of their dams and lodges, and where they can find aspens, willows, and birches; the bark and buds of these trees are their favorite food. A beaver grows to a length of four feet and may weigh as much as sixty pounds. As we have noted when describing the feeding adaptations of herbivores, beavers possess incisor teeth well suited for stripping bark and cutting down trees. A beaver is one of our most amphibious freshwater mammals, equipped with webbed hind feet for swimming, a flattened hairless tail that serves as a rudder and sculling oar, and an ability to hold its breath underwater for as long as fifteen minutes. No other mammal has played such a significant role in early American history. It was a search for beaver pelts which sent many white men into the remote parts of the American wilderness. In pioneer days the skins were a medium of exchange, and many American fortunes were built on the exploitation of this mammal. In one twenty-five-year period of the last century, the Hudson's Bay Company alone brought three million skins to the London market. The resulting impact upon the beaver population led to its near extinction. Fortunately a decline in trapping, with an introduction of modern conservation and protection policies, averted this fate. Today the beaver is holding its own and, in many protected wildlife sanctuaries, actually increasing in number. Its role as an engineer in flood control and dam construction, its ingenuity in building lodges make the beaver a valuable ally of man as well as a fascinating member of the wildlife community. We can be thankful the beaver did not go the way of the passenger pigeon and the bison.

THE LIVING ENVIRONMENT: THE CARNIVORES

As in all biomes, the herbivorous animals of the deciduous forest are near the base of the pyramid of life. Whether insect, bird, or mammal, they convert the energy produced by the green plants into available animal energy for all the other forest animals. At the apex of the wildlife community are the carnivores. Some of the carnivorous beetles, bugs, and bloodsucking insects of the deciduous forest are like those we found also in the conifer biome. The higher consumer niches of both biomes are filled by vertebrates; among these, the cold-blooded amphibians and reptiles of the deciduous forest are more numerous than in the boreal forest. The warm-blooded vertebrates of the consumer niche are well represented by the birds, which are unusually abundant in this deciduous biome. Today the mammal carnivores are relatively rare because of the destruction of the original forest and the persecution of predators by man.

The humid climate, the warm summers, and the many available aquatic habitats of the deciduous forest biome provide an ideal environment for amphibians. No other biome includes such a great variety and abundance of frogs and salamanders; the deciduous forest possesses more species of salamanders than are found anywhere else in the world. The majority of eastern frogs and salamanders are restricted to the deciduous forest biome, though a few species range north into the conifer biome and westward into the grasslands.

Frogs and toads have become adapted to a variety of habitats; many live in or near the water, others are arboreal and adapted for life in the trees, and a few have chosen to live in the shade and protection of land vegetation. Most aquatic are the leopard frog and bullfrog. The LEOPARD FROG, almost transcontinental in range, is the common meadow frog that splashes into a pond at our approach. It has rounded black spots on a well-camouflaged yellowish green body. The leopard frog fills the insectivore niche from Labrador south to the Gulf of Mexico and west to the Rocky Mountains. Largest and most aquatic of all is the BULLFROG, a mottled green and brown frog reaching a length of six inches or more, common in all river swamps and marshy ponds. Bullfrogs have a wide range, living in the boreal forest and on the prairies as well as in the deciduous forest biome. They are especially numerous in the lower Mississippi River lowlands, where they are caught and marketed for the flavorsome meat of their hind legs. All frogs rely almost entirely upon insect food, spearing their tiny prey on sticky tongues that dart out with lightning speed.

Smallest of the frogs are the treefrogs, one inch or less in length. They are exceedingly vocal members of the forest community after every spring rain. Treefrogs have long legs and prehensile toes tipped with adhesive discs. Thus they are adapted for an arboreal life; they feed on the many insects that infest foliage. The GREEN TREEFROG, emerald-green in color, is a tiny forest sprite found chiefly in the Louisianian life zone. The southern oak-pine forests often ring with the deafening chorus of thousands of these diminutive vocalists after a spring rain. The common SPRING PEEPER, a grayish-brown species found throughout the eastern United States, is a familiar harbinger of spring in the deciduous forest when these treefrogs make swamps and ponds resound with their bell-like tones.

SPOTTED SALAMANDER

MARBLED SALAMANDER

Most salamanders live in moist-land habitats, near brooks and ponds. They feed on worms, larvae, and ground-dwelling invertebrates. Of the many species found in the Appalachian Highlands and adjacent lowlands, a few are typical of the group; we can probably discover these if we take time to overturn some stones or look under a log. The most common salamander of the northeastern United States is the RED-SPOTTED NEWT. Newts have an unusual life history; the adults live an aquatic existence, but the young spend several years living on land. The adult newt, three to four inches long, normally is olive-green with red spots. The young newts are bright red and also spotted with red; they can be found hiding under stones and leafy litter. They are often mistakenly considered a separate kind of salamander, but this "red eft" form is merely a land stage of the red-spotted newt before it becomes the aquatic adult. The DUSKY SALAMANDER, the same size as a newt, is the common dark brown, slimy salamander found beneath stones and logs; it is most abundant in the Appalachian Highlands. The SPOTTED SALAMANDER, a glistening brownish black, is marked with two rows of large yellow or orange spots along the back; it is found from Nova Scotia to Louisiana.

The deciduous forest biome includes two of the largest amphibians of the continent: the mudpuppy and the hellbender. The MUDPUPPY is an eel-like grayish-brown animal growing to a length of thirteen inches. It has very short legs and feathery external gills, which form red plume-like appendages on each side of the head. Mudpuppies lead a nocturnal life, inhabiting streams and ponds and feeding on crayfish, small fishes, mollusks, and insects. We find these amphibians throughout the biome except in northern New England and on the eastern slopes of the Appalachians. Giant of all American amphibians is the HELLBENDER, a grotesque salamander which grows to be twenty inches in length. It usually causes consternation among fishermen who hook it by mistake. The head is broad and flattened, the mouth huge, and a frill of loose skin forms a fold along each side of the body. This salamander, harmless though large and hideous, lives in streams and ponds of the western Appalachian Highlands, from Kentucky to the Ozark Mountains; it feeds on aquatic insects, crayfish, and worms.

In neither of the two previous biomes have we found many reptiles. But in the milder climate of the deciduous forest biome, we encounter a number of different turtles, lizards, and snakes — all occupying carnivore niches in the food chains. Active during the summer season, dur-

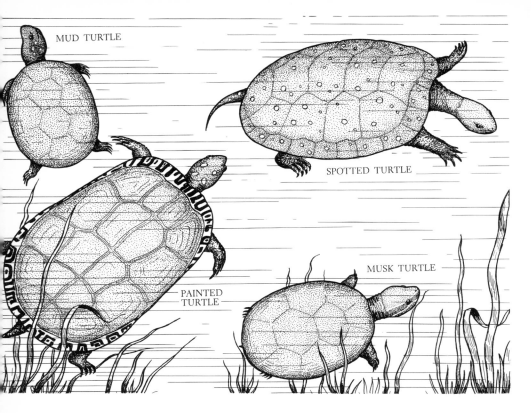

ing the cold months they become dormant, buried beneath the ground or in the mud at the bottom of ponds.

Turtles are common reptiles everywhere in the biome, being adapted for living both in the water and on land. Several species have a wide range throughout the entire deciduous forest, wherever there are ponds or swampy lowlands. In this group we find the aquatic painted snapping turtles. The PAINTED TURTLE is found from coast to coast, although only in disjunct colonies in the south-central and southwestern states. We often see a painted turtle basking on a log or rock, its flat olive-brown shell edged with a colorful margin of red, yellow, and black. Painted turtles feed on insects, crayfish, and small mollusks found in the water. Largest of the freshwater turtles is the common SNAPPING TURTLE, found in muddy streams and marshy ponds throughout the eastern United States. Snapping turtles grow to be fifteen inches long and thirty pounds or more in weight. The mud-colored shell, often camouflaged with a growth of algae, has three lengthwise keels; the long fleshy tail is armed with spines at the tip. This ugly and aggressive reptile hides in the bottom debris, with head protruding and jaws ready to snap at any aquatic animal that comes within reach. One is wise to keep a safe distance from the head of this quick-tempered reptile;

it can strike with great speed and crush a hand in its powerful jaws.

In the drier parts of the forest we may discover some of the more terrestrial species of turtles. The common EASTERN BOX TURTLE ambles slowly through grass and underbrush, its yellowish domed shell inconspicuous among its surroundings. The young are carnivorous, feeding on larvae and worms; adults, however, are herbivorous, preferring berries, fruit, and foliage. The box turtle has discovered the ultimate in self-protection. When disturbed, it can pull in head and limbs, close the hinged lower shell, and wait safely in its armored box until the curious predator gives up and goes away. This mild-mannered member of the deciduous forest holds the record for longevity (more than 130 years) among all American animals. Most terrestrial of all the turtles is the GOPHER TORTOISE found in sandy woods of the southern coastal forest; it is most likely to be encountered in Florida. The grayish sculptured shell is as distinctive as the pillar-like limbs, with toes adapted for digging. Its burrows are often a foot in diameter and extend for thirty feet underground. This turtle's feeding habits are an exception to those of most reptiles, for it is a strict herbivore, feeding on grasses, foliage, and fruits.

Lizards are not as abundant in the deciduous forest as in the more arid habitats of the western United States. However, any exploration of the Carolinian and Louisianian life zones of this biome is likely to result in meeting one of these small but interesting reptiles. All eastern lizards are carnivorous, being particularly fond of flies, ants, and other small insects. Two types of lizard are common members of the deciduous forest community: fence lizards and anoles. Fence lizards, or swifts, have rough pointed scales and are camouflaged in mottled gray and brown, which makes them practically invisible on the bark of trees. The common FENCE LIZARD, about six inches long, is an arboreal reptile, adept at scurrying up tree trunks. It can often be heard rustling through fallen leaves when looking for crickets, spiders, and millipedes. The GREEN ANOLE, or "chameleon," is undoubtedly the characteristic lizard of the Louisianian life zone. This agile and attractive little lizard has special climbing pads on its toes, and can cling to vertical surfaces with ease; it is usually found on the trunk or branches of trees, and even on the underside of the leaves. Green anoles become inactive when the temperature falls below 70° F., yet cannot stand direct exposure to the sun at higher temperatures. They are continually alert for flies, mosquitoes, caterpillars, and moths. An anole has the unusual ability to change color, the green anole turning brown under certain conditions, bright green at other times. In general, bright light or low temperature will bring on the dark coloring; in shade or at high temperatures, the green color appears, and also when the creature is excited or disturbed. The color change is not correlated with the color of the environment and therefore is not primarily a protective adaptation.

Fear of snakes undoubtedly prevents many people from thoroughly enjoying the wilderness, but most snakes are harmless and will try to avoid anyone they might meet on a woodland trail, being more eager to escape detection than to attack. Although a few species are poisonous, these are fortunately in the minority and restricted to special types of habitat. Snakes are as important in the "balance of nature" as any other group of animals. All snakes are carnivores, though they rely less upon insects as food than do the smaller amphibians and lizards; many prefer warm-blooded animals as prey, and are a vital check on overpopulation of the community by rodents. Snakes are adapted to a variety of habitats in the forest. The majority are terrestrial, but some are aquatic and others arboreal. Although many different kinds of harmless snakes live in this biome, generally they escape notice because of their secretive habits.

Probably the first snake you encounter when you walk in the woods will be an EASTERN GARTER SNAKE, which occurs throughout this biome and also in the boreal forest. It is a slender snake, rarely more than three feet long, with lengthwise yellow and black stripes. It hides under rock piles and fallen trees, lying in wait for toads, earthworms, or similar small prey. The common EASTERN MILK SNAKE, or "checkered adder," grows slightly larger than the garter snake; it is gray or tan, with a pattern of darker blotches on the back. It is a constrictor, prowling the woods and fields by night in search of mice and other rodents. When approached it may coil and strike, but its bite is harmless. The much larger BLACK RACER, or blacksnake, grows to a length of five feet. This is a blue-black snake capable of traveling with great speed over the ground and through the branches; it climbs trees readily, and thus can reach birds' nests and rob them of the eggs which are a special treat. The black racer also eats rodents, amphibians, and smaller reptiles.

You may meet some poisonous snakes in the deciduous forest, since this biome is the home of five species with fangs and venom sacs. It is well to be able to distinguish these from their harmless relatives. Most widespread is the TIMBER RATTLESNAKE found in all the Appalachian

Two poisonous snakes of the deciduous forest are the copperhead (upper) and timber rattlesnake (lower). (*Credit: Courtesy of The American Museum of Natural History*)

Highlands except in Maine. The favorite haunt of this snake is in dry rocky woods where ledges for sunning are available and crevices for dens. A timber rattlesnake is yellowish or brown, marked by saddles of darker brown or black. It is a thick-bodied snake, usually under four feet in length, with a triangular head; the rattle terminating its tail is the identification badge of the rattlesnake group. Much larger and far more dangerous is the EASTERN DIAMONDBACK RATTLESNAKE which grows to a length of six feet or more; it is the largest of all the rattlesnakes found in the United States. Its range is usually only on high dry ground in the oak-pine woods of the coastal forest; it is rarely found north of the Carolinas. The color of the eastern diamondback is olive or brown, marked with a bold pattern of diamond-shaped spots on the back. This coloring and marking so completely camouflage the snake, as it lies in patches of light and shadow on the forest floor, that you can walk within a few feet of the snake without seeing it. Its food consists of rodents, rabbits, and small birds. Strangely enough, one of its chief enemies is another snake, the kingsnake, which is immune to rattlesnake venom.

In open rocky woods in the northern part of the biome lives the NORTHERN COPPERHEAD. The copperhead has a velvety-brown body marked by darker brown crossbands, and a copper-colored head. Like the rattlesnakes, it is a stocky snake, although rarely more than three feet in length. It feeds on shrews, mice, caterpillars, and large insects. In a quite different habitat lives the COTTONMOUTH, or water moccasin, whose home is in the southern cypress swamps and marshy lakes from Virginia to Texas. Most cottonmouths are dingy brown in color and less than four feet in length. Their diet consists of cold-blooded prey, especially frogs and catfish. These snakes are known as cottonmouths because their mouth when open reveals a distinctive white, cottony throat. Smallest of the eastern venomous snakes is the EASTERN CORAL SNAKE, a New World relative of the cobra, with equally poisonous venom. Fortunately its brilliant coloring is easily recognized, a series of broad bands of black, scarlet, and yellow. The eastern coral snake is a slim snake, usually under two feet in length; the fangs are no larger than rose thorns, unable to penetrate clothing. Hiding under boards and debris, this small burrowing snake finds lizards, frogs, and even smaller snakes for its meals. This species is common only in Florida and adjacent states.

The bird life of the deciduous forest is so incredibly abundant that it is difficult to do it justice in a few words. Not only is the entire biome an ideal environment for many permanent residents, but much of the Appalachian Highlands lies in the main flyway followed by bird migrants in their annual north-south travels. A great majority of these are the omnivorous and carnivorous species that occupy the uppermost consumer levels in the food chains. These bird members of the wildlife community present a vivid ecological picture of the extent to which specialization can take place in the distribution of animals through different niches. Even in a brief exploration of the forest, we soon discover that the bird species are not scattered throughout the various habitats in a hit-or-miss fashion, but that each has its own particular niche in its preferred environment. A representative walk through the biome — for example, in Shenandoah or Great Smoky Mountains National Parks — can take us through four types of habitat, each with a diversity of niches, where we see four types of birds: (1) the forest dwellers, being found among the trees of the dense woods; (2) those adapted for life in the open, in the grassy glades and cutover thickets of the forest edge; (3) a few inhabitants of woods and fields bordering on lakes and streams; and (4) the birds of prey.

Forest Dwellers. These include the many small birds whose cheery songs echo through the dense woods and add so much to any walk through the deciduous forest. The majority have a wide range and are found throughout the biome. They rely chiefly upon insects for food, but when these are not available they live on seeds, fruits, and other parts of plants. Though all have the same general food preferences, they do not compete with each other because of their selection of individual niches. Some feed on the ground beneath the trees, others on the trunks of trees, and still others among foliage.

On the ground, scratching among the fallen leaves and litter of the forest floor, are the birds that search for insects and their larvae, worms, and other ground-dwelling invertebrates. Typical of the birds in this niche are the wood thrush, ovenbird, towhee, and brown thrasher.

The trunks of the trees constitute another niche for insect-eating birds. The many bark-infesting insects and their larvae provide rich feasts for the birds whose specialized feet and bills enable them to occupy this niche. Among these we find nuthatches, the brown creeper, and numerous woodpeckers. Nuthatches have a peculiar habit of creeping down a tree trunk headfirst as they pry beneath the bark for insects. The WHITE-BREASTED NUTHATCH feeds on nuts and acorns in addition to insects. It has learned to wedge a seed in a crevice of the bark for

202 / THE BIOMES OF NORTH AMERICA

storage, then to hammer (or "hatchet") out the kernel with its strong, sharply pointed bill. The BROWN CREEPER spirals up a tree trunk looking for insects on the way, then flies to a different tree and drops to its base; next, the whole process is repeated. The brown creeper has an unusually slim, long, curved bill that can reach into bark crevices for insects and caterpillars. Like the red-breasted nuthatch, this insect-eater is also more common in the northern part of the biome. Noisiest of the bark feeders are the woodpeckers, which go at their business of dislodging insects with resounding vigor. Two members of the family which we met in the conifer biome occur also in the deciduous forest, the hairy and the pileated woodpeckers. In addition we may find the largest and rarest of the woodpeckers, the IVORY-BILLED WOODPECKER, with a length of twenty inches. It closely resembles the pileated woodpecker, and has a similar large red crest; its bill, however, is ivory-colored instead of black. Formerly this species occurred from the Carolinas to Texas, but today is on the verge of extinction as a result of being hunted for food and because of destruction of its habitat. Few living naturalists have seen this rare bird, whose existence has not been reported in the last decade.

Many insect-eaters spend all their lives in the trees, different species choosing their own niche, some near the ground, others nearer the treetops. Two birds we commonly see in the lower branches, assiduously examining every branch and leaf for insects, are the BLACK-CAPPED

THRASHER   WOOD THRUSH   MEADOWLARK

CHICKADEE and the TUFTED TITMOUSE, the latter a gray bird the size of a large sparrow, with a darker gray crest. Other birds use the trees only as a perch from which to take off and catch flying insects. Such is the habit of the AMERICAN REDSTART, a black bird with conspicuous orange markings on wings and tail. It is the jumping jack of the arboreal birds, constantly taking to the air to intercept a flying insect.

Dwellers of Open Woods and Fields. With the destruction of the original dense forest cover, an increasing number of open habitats have appeared: grassy clearings, thickets, cutover woodlands, and abandoned farms. Here are other niches in which we find different species of birds from those in the denser forests. Typical of the birds in this niche are the bluebird, goldfinch, eastern meadowlark, and yellow-shafted flicker. Few birds are as familiar to the roadside naturalist of this biome as the colorful EASTERN BLUEBIRD. This cousin of the thrushes has a bright blue back, reddish-brown underside. It is a bird of the open country, feeding on grasshoppers, beetles, and caterpillars in season, and subsisting on berries of sumac, poison ivy, and bayberry in winter. The AMERICAN GOLDFINCH, or wild canary, is another cheery bird of the open fields; its plumage is a golden yellow, with contrasting black wings and face. Goldfinches fly with a characteristic swooping movement; they too fatten on insects in summer, but must rely upon weed seeds and fruits of birch and alder in winter. Two larger birds occupy niches in the open grassy areas of the biome. The EASTERN MEADOWLARK is a brown-

DOWNY WOODPECKER

RED-HEADED WOODPECKER

IVORY-BILLED WOODPECKER

FLICKER

spotted bird with conspicuous yellow breast and black bib beneath the throat. The meadowlark flies low, catching beetles, grasshoppers, and other insects of the fields with its long pointed bill. This bird seems to enjoy pouring forth its melodious, whistling song as it sits in a treetop or on some elevated perch. The YELLOW-SHAFTED FLICKER, a woodpecker that has deserted the trees for the ground, inhabits woodlots and edges of open woods. This large ground-feeding bird has a speckled brown body with yellow underwings and a red crescent on the back of its head. Though often seen on the ground in company with robins, it searches for ants rather than earthworms. In winter flickers give up their insect diet for one of fruits and berries.

Dwellers of Swamps and Pond Margins. The borders of ponds and swamps provide niches for two other familiar birds of this biome, the kingfisher and the red-winged blackbird. The BELTED KINGFISHER often sits on a limb above the water, looking into its depths for a favorite meal of fish. When a fish is detected the bird dives like a feathered missile and rarely misses its prey. Kingfishers are built like prizefighters, with great strength concentrated in their muscular neck and chest. The plumage is grayish blue, the oversized head is capped by a shaggy and unkempt crest, the bill is unusually long and straight. The RED-WINGED BLACKBIRD is the inevitable frequenter of swaying cattails and clumps of alder. This swamp dweller is a shiny black bird with small but striking wing patches of red and yellow. Redwings feed on insects, can-

RED-WINGED BLACKBIRD

GREAT BLUE HERON

KINGFISHER

kerworms, and caterpillars; in winter they subsist on seeds and grains. Wooded swamps are also the home of many different kinds of herons, the long-legged wading birds which feed on aquatic animals such as fishes, frogs, and snakes. The woods offer few more arresting sights than that of a GREAT BLUE HERON poised motionless in shallow water, a four-foot statue of grace and beauty. The blue-gray plumage has warm brown markings on the wings, the head is white, with backward-reaching black plumes, the yellow bill long and sharp. In capturing aquatic prey, the heron strikes swiftly, spearing the victim and tossing it into the air; then the dexterous bird catches the meal deftly in open mouth.

Birds of Prey. These are the top carnivores of the bird world in the deciduous as in the conifer forest. Rarely confined to any special habitat, they range widely wherever they can find suitable food. The niche of the daytime hunters is taken over by the hawks and eagles, of the nocturnal hunters by the owls. Wheeling and soaring in the sky, scanning the earth beneath with sharp eyes, are the red-tailed and red-shouldered hawks. Also gliding aloft are other birds of prey with a wide range: the osprey, bald eagle, and the black vulture. The RED-TAILED HAWK is a huge predator with four-foot wingspread, capable of sustained soaring flight at high altitudes. Like most hawks it is an important check on the rodent population, since its preferred diet consists of mice. Amphibians, snakes, lizards, rabbits, and shrews also make up its fare in the eastern part of the biome; in the midwestern states, gophers and grass-

SCREECH OWL

BLACK VULTURE

TURKEY VULTURE

hoppers are the main food items. The RED-SHOULDERED HAWK is slightly smaller, with a three-foot wingspread. This is a more sluggish hawk, most frequently seen in swamps and bottomland forests, where it feeds on cold-blooded prey — chiefly frogs, crayfish, and insects.

Vultures are large carnivores, although their feet and claws are not as strong as those of the hawks. They are therefore scavengers feeding on weak, disabled, or dead animals. By feeding on carrion, these birds perform a valuable service in cleaning the highways, where so many wild animals meet sudden death. Vultures have a wingspread of five feet or more; on the ground they hobble clumsily, but in the air are graceful fliers capable of soaring effortlessly at great heights. The TURKEY VULTURE, or "buzzard," is common in the southern part of the biome. As this big black bird with naked red head and hunched shoulders perches on a treetop, it presents an ominous and ghoulish appearance. A group of circling turkey vultures is a sign that one of their kind has discovered a dead animal below; soon the vultures congregate around the carcass and the feast is on. The BLACK VULTURE is a smaller bird, with feathered head and neck and a shorter square-cut tail. It does not range far west of the Mississippi River and is most abundant in the southeastern states.

The niche of the nighttime hunters is occupied by the owls. Almost as large as the great horned owl of the boreal forest is the BARRED OWL, a dweller in dense woods. Though not as aggressive, it fills its own carnivore niche in the community. This is a large grayish-brown owl, lacking the eartufts of the great horned owl. Its diet covers a wide choice from mice and other small mammals to frogs, birds, and large insects. When you are out in the evening, you are likely to discover the little SCREECH OWL. Common around dwellings, this owl has grayish or brownish plumage and small eartufts. Screech owls eat any prey they can overpower; like the other owls it includes a great variety of animals in its diet, from insects and crayfish to birds and small mammals.

The larger mammal carnivores are either very rare, or have entirely disappeared. The black bear and mountain lion, which still inhabit the conifer biome, were formerly abundant in the deciduous forest biome also. But they have had to retreat to the few mountain wildernesses in the national parks and forests of the eastern United States. Black bears are seen occasionally in the Appalachian Highlands and in Florida, but it is newsworthy for a hiker to come across one of these large mammals even there. Of the smaller carnivores that we met in the boreal forest, some also occur in the deciduous biome and serve as useful predators; among these are several species of weasel. The typical carnivores of this biome are small or medium-sized mammals. Most are nocturnal in habit and so are infrequently seen on daytime explorations of the deciduous forest. Three of these are arboreal in their adaptations and often are found in the trees: the raccoon, opossum, and bobcat. Two are more commonly restricted to the ground: the striped skunk and the red fox.

The RACCOON is a medium-sized carnivore reaching a weight of thirty pounds or more. It is a grayish animal with black-banded tail and face distinctively marked by a black mask. Raccoons are plucky fighters with sharp claws and teeth; thus they have few natural enemies. They are also crafty and intelligent animals, and have managed to survive in all the eastern states. Their natural habitat is a wooded swamp near streams which provide their favorite food of crayfish and other small crustaceans and frogs and fishes. They also feed on insects and relish such plant food as nuts, berries, and fruits. Raccoons have become semitame animals in parks and recreation areas, where, as daytime visitors, they beg for food and provide entertainment with their amusing antics. Even though hunted for food and for their pelts, raccoons are increasing in numbers.

Second-growth thickets, which abound in many parts of the biome today, offer an ideal habitat for the STRIPED SKUNK, also seemingly in little danger of becoming extinct. This small carnivore, of house cat size, is found throughout the United States. Its markings are very distinctive: the fur is black, with contrasting white forehead and two lengthwise white stripes along the back. Skunks avoid deep forests and, being omnivores, often invade farmers' fields for berries, fruits, and corn. Their animal food consists of insects, rodents, and amphibians. Skunks are most likely to be encountered at dusk or after dark. This mammal is notorious for the perfection of its own type of chemical warfare. A pair of musk glands beneath the tail can direct a well-aimed spray of acrid, odorous fluid which keeps most members of the wildlife community (and man!) at a respectful distance. The only predator that ignores the scent seems to be the great horned owl, a skunk's mortal enemy.

An OPOSSUM, a relic of the most ancient order of mammals, is far from a beautiful animal by any standard. It has a yellowish-white, ratlike, and pointed face, naked black ears, and a large mouth that opens in a snarling expression to reveal many small teeth. The naked tail, pre-

The opossum is a relic of an ancient order of mammals; it lives in the deciduous forest. (*Credit: Ross Allen*)

hensile like that of a monkey, serves the animal as a fifth leg when climbing among tree branches. Its unkempt fur coat consists of stringy guard hairs, some black and some white. Opossums are unusually agile climbers, aided by opposable big toes on the hind feet. They are most active at night, when they forage for grasshoppers, crickets, worms, mice, and amphibians. Their original range was restricted to warmer parts of the biome from Virginia to Texas, but they have recently invaded the Carolinian life zone and are becoming plentiful as far north as northern New York State. This may be the result of a long period of increasingly mild climate, since this mammal cannot survive cold winters. The opossum has the strange habit of appearing dead — "playing 'possum" — when attacked. The excitement of being pursued throws the animal into a state of nervous shock, causing it to become paralyzed with fear; it then falls in a limp heap, eyes closed and tongue hanging out. When danger is over, the 'possum recovers quickly and makes a fast getaway. This behavior undoubtedly has some survival value if the pursuer is a carnivore that refuses to eat a dead animal.

The RED FOX is one member of the wildlife community which has eagerly accepted the challenge of living dangerously — close to man. This remarkably intelligent and adaptable carnivore, although essentially nocturnal in habit, is often seen by day trotting across the fields like a small reddish-brown collie dog with a very bushy tail. Foxes rarely stay in dense forest, but prefer open woods and fields. Here they can find an ample supply of rabbits, mice, woodchucks, and chipmunks. They also eat birds, reptiles, insects, and carrion and are therefore a valuable check on the otherwise increasing number of small herbivores.

Top carnivore of the deciduous forest today is the BOBCAT, or wildcat. A bobcat has buff or brown fur streaked with black, and weighs as much as twenty-five pounds. It is a close relative of the Canada lynx, which it closely resembles except for smaller eartufts and shorter, less clumsy limbs; the bobbed tail, like that of the Canada lynx, is responsible for its name. Bobcats occur throughout the United States, hiding in forested swamps and inaccessible wilderness areas. Seldom seen by day,

The fox is an intelligent and adaptable carnivore that often lives close to farms and towns. (*Credit: New York Zoological Society*)

this predator is very successful in keeping out of sight, fading silently into the shadows of the woods before it can be detected. Bobcats climb with ease and swim when they have to. At night they hunt for hares and rabbits; in winter, when smaller mammals are scarce, they will tackle a full-grown deer. Being less well adapted for traveling through snow, bobcats do not range as far as lynx. A bobcat is a proverbially fierce opponent when cornered, and today has few natural enemies but man. This valuable predator has been greatly reduced in number by the lure of bounties which encourage its indiscriminate killing.

Man and the Deciduous Biome

The climatic and topographic features of this biome are favorable for agriculture and industry; to make way for these necessary adjuncts of civilization, much of the original forest had to be destroyed. Added to these circumstances was the usefulness of the deciduous trees for lumber. Together, all these conditions made inevitable the destruction of the virgin forest. Today we rarely find a survivor of the original forest, such as a 300-year-old white pine or a 500-year-old oak. Gone too, with the disappearance of the dense forests, are the large mammals that once inhabited them.

The change from forests to cutover woodlands and second-growth thickets has caused many changes in the wildlife as well. The new type of habitat encouraged the spread of animals adapted to live in the newly created niches. Many herbivores have increased, such as rabbits and deer; with them have come added numbers of omnivores and carnivores like the red fox and raccoon. With the introduction of many exotic species of ornamental plants, especially around suburban houses, a new and greater population of plant-feeding insects has arisen; this has led to a proportional increase in the number of insect-eating birds. Over the entire biome those animals have survived and even increased in number which have been able to adapt themselves to competition with man and his enterprises.

The forests of the southern Appalachian Highlands have been least modified by man. In the higher, more rugged elevations of Appalachia some primitive areas have been preserved in Shenandoah and Great Smoky National Parks. Others have survived in the Green Mountain and White Mountain National Forests, in such state parks as Adirondack State Park of New York and Baxter State Park in Maine. The lowland forests, especially west of the central highlands, have suffered the greatest change and will undoubtedly never return to their original forested condition. The coastal-plain forest is more adequately protected; here the presence of large and impenetrable swamps has halted the march of civilization. Foresighted management of the pine forests also is preserving some of the original environment of this part of the biome.

The deciduous forest biome may not present as spectacular scenery as is found in many parts of the conifer biome, and it may not include as many large and impressive animals. It does, however, present a unique variety of species in an almost infinite diversity of habitats and niches. Adaptation, competition, survival, succession, food chains — all these are as evident within a few miles of New York City or Chicago as they are in the more inaccessible Rocky Mountain forests. The deciduous forest is a living laboratory, accessible and easy to explore, wherein we can see many of the fundamental ecological processes at work. If we will take time to know these neighborhood woods and fields, we shall find in them fascinating and instructive demonstrations of the biotic community in action.

11 / At the Edge of the Tropics

THE TAMIAMI TRAIL crosses the peninsula of Florida from the Atlantic Ocean to the Gulf of Mexico. To many it may seem merely a shortcut across the watery wilderness from Miami on the populous east coast to Fort Myers on the west, a monotonous 150 miles through a wasteland of marshes, grasslands, and swamp forests. But to an ecologist it is an intriguing route along the edge of the tropics through wildlife communities found nowhere else on the continent. South of the Tamiami Trail, following the Overseas Highway, we traverse the Florida Keys, a 200-mile chain of islands extending southward from the tip of the Florida peninsula. Here we are at the edge of the tropical biome.

As the Tamiami Trail leaves Miami, it passes through pine groves interspersed with grassland and palmetto thickets much like the vegetation of the coastal plain to the north. Soon, however, it leads through a seemingly endless sea of tall grasses; for fifty miles we drive through that vast freshwater marsh known as the Everglades. Startled by our approach, flocks of white egrets take to the air from the waterways; huge turtles slide off logs into water teeming with fishes; basking alligators slip into floating beds of water hyacinth. Leaving the 'Glades we enter Big Cypress Swamp, an open forest of low-growing cypress trees studded with bristling clumps of air plants. Dome-shaped "islands" of angiosperm trees, called hammocks, rise above the level swampland. Each hammock is a dense growth of trees, shrubs, and swinging vines; aerial gardens of air plants and orchids cling to the branches. On the Gulf Coast the Tamiami Trail skirts mangrove forests, into whose protection have retreated two of our rarest animals: the crocodile and the manatee. In this short span of a few hours' drive, we have passed through the four main types of communities we shall explore at this edge of the tropical biome: the Everglades, the cypress swamps, the hammocks, and the mangrove forests. Let us continue on foot and by boat to get a closer look at the environment of the unusual region, and at the plants and animals adapted for living in it.

THE TROPICAL FOREST BIOME

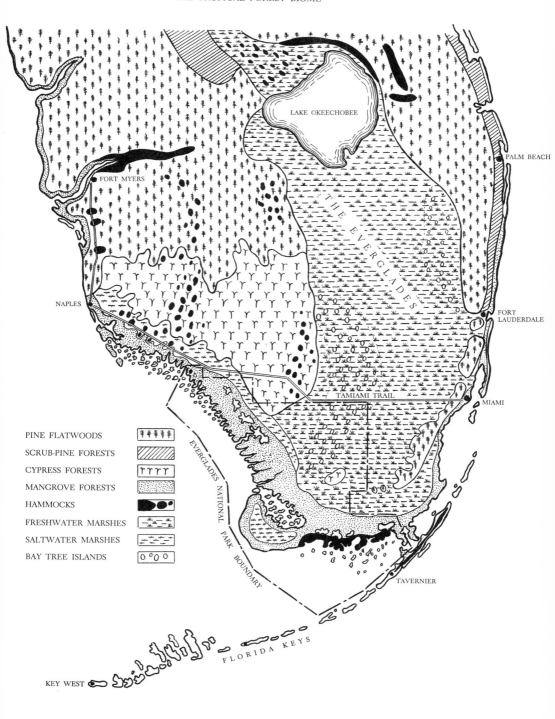

The Physical Environment

If the peninsula of Florida did not extend to within a few degrees of the Tropic of Cancer, there would be no tropical biome in the United States. Since it does, this southern part of the state provides a suitable home into which have migrated many tropical plants and animals characteristic of the West Indies, Central America, and South America. North of the Tamiami Trail, the wildlife is a mixture of species typical of temperate regions, with others indigenous to the tropics.

Surrounded on three sides by the tempering influence of the ocean, this portion of Florida, where subtropical and tropical conditions merge, has a climate favorable to a lush growth of warm-climate plants. Temperatures vary little from summer to winter. The January isotherm of 64° F. extends across the peninsula from Palm Beach to Fort Myers, forming an approximate northern limit to the biome. The average July temperature is between 80° and 84° F., and so there is a difference of less than twenty degrees between average winter and summer temperatures. Although frosts do occur occasionally over most of the region, they are infrequent. The growing season is at least 300 days; in the tropical area no frost interrupts a growing season which lasts for the entire year.

Humidity is high throughout the year. Rainfall is abundant, varying from fifty to sixty-five inches annually; the heaviest precipitation occurs in the coastal Miami area, which is often drenched by torrential downpours. This rainfall is not evenly divided over the year. The six-month period from November to April is relatively dry, with two inches or less of rain per month. On the other hand, the six months from May to October are very wet, with as much as nine inches of rain in some months. This creates seasonal floods and accounts for the type of vegetation found in many of the wildlife communities. As a result the majority of plants are either mesophytes or hydrophytes, and water-dwelling vertebrates (fishes, amphibians, reptiles, water birds) are abundant.

In no other biome is the land surface so uniformly flat; elevations are for the most part only a few feet above sea level. High ground in the Everglades and cypress forests is only fifteen feet above sea level; a coastal dune forty feet is a commanding hilltop. Thus topography has little effect on the type of wildlife community. Far more important factors are drainage, soils, and the alternation of wet and dry seasons. Grasses and other herbaceous vegetation can obtain sufficient water from the surface soil at all seasons, but trees require a deeper source of groundwater to sustain them through the dry season. Water levels in the ground are therefore responsible for the presence or absence of trees. The scrub cypress forests grow on shallow soil underlain by impervious rock layers. Only those trees which can find pockets in the subsurface limestone layers are able to obtain adequate water in the dry season to maintain normal tree growth. Since the underlying rocks in the Big Cypress Swamp have few such pockets, the trees are of small size and widely scattered. On deeper, moister soils of hammocks, with ample and continuous groundwater supply, the trees of this biome reach their most luxuriant development.

Lake Okeechobee, at the northern edge of this region, is an unusual inland body of water, extremely shallow, yet covering 700 square miles. Choked with aquatic plants, especially floating water hyacinth, it resembles a swampy plain more than a lake. On either side of Lake Okeechobee, porous sandy soils extend into southern Florida, and support pine forests with an undergrowth of grasses and shrubs. Along the east coast other sandy areas create the beaches, offshore bars, and dunes where many xerophytic and salt-tolerant plants live. South of Okeechobee are the freshwater Everglades marshes, actually a slow-moving river forty miles wide and ninety miles long, draining into the Gulf of Mexico. Deep peat and muck soils underlie the Okeechobee-Everglades basin, which constitutes one third of the entire area of this biome. Beneath the Everglades are impervious rock layers that permit little seepage; as a result the 'Glades are flooded in the wet season, become a dry grassland during the rainless winter. The southern tip of the peninsula is a marl-muck area where there are few sand beaches, many mudflats and tidal marshes. This region is flooded with each rise of the tide; although the four-foot tide range is slight compared to that of the northern seacoasts, extensive inland areas are inundated because of the flat topography. Here we find the mangrove forests.

THE LIVING ENVIRONMENT: THE PLANT LIFE

The change from the deciduous forest landscape to that of southern Florida takes place gradually. We leave behind the white oaks, beeches, and tuliptrees as we enter Florida. But we still can find the ubiquitous red maple and bald cypress in the swamps; flowering dogwood, redbud, and holly among the woods of oaks and pines. Then, one by one, new tropical species appear: royal palms, strangler figs, epiphytic orchids. The vegetation of southern Florida includes ninety kinds of trees, prob-

Cypress forest mingle with the saw grass in the western part of the Everglades. Clusters of epiphytes resemble birds' nests on the cypress trunks. Everglades National Park, Florida.

ably a greater number than found in any other American area of comparable size except in the Great Smoky Mountains. The great variety of vegetation is brought about by the overlapping of temperate, subtropical, and tropical species.

The plant life, as in the other biomes we have explored, consists of a number of smaller communities, or associations, each of which develops under special soil and drainage conditions. The major types of association we shall explore are those of the Everglades, the pine forests, the cypress swamps, the hammocks, and the mangrove forests. In the *Everglades*, herbaceous plants dominate the community, many square miles being covered with a rough-stemmed sedge known as saw grass. On hummocks in the midst of level stretches of saw grass, grow marsh and royal ferns. Low depressions are filled with cattails and rushes; open canals with gardens of pond lilies and pickerelweed. *Pine forests*, known as pine flatwoods, occupy the sandy higher ground. Slash pine, the dominant tree, forms open groves through which we gain vistas of grassland, stretching to the level horizon. Amid the grasses grow clumps of palmetto, and cabbage palms force their way through the jungle undergrowth. This slash pine–palmetto–cabbage palm community is typical of the northern part of the biome. A more open woodland of sand pine and scrub oak is found along the eastern coast. Here the ground is dry and sandy, and few plants thrive but saw palmetto and prickly pear cactus. A surprising member of this community is reindeer lichen, forming gray crisp cushions on the sterile ground; it is a long way from the tundra! *Cypress forests* occur along stream and pond borders in the pine forests; they also grow west of the Everglades marshes to the Gulf Coast. Surrounded by their aerating knees, old cypress trees often stand in water several feet deep. Young cypress growth, however, will not develop where water covers the ground all year, since the seedlings require atmospheric oxygen.

Mangrove forests occur on low-lying tropical shores throughout the world. They are unique, because few trees are adapted for living in the sea. Yet mangroves not only can grow in saltwater but are able to withstand violent wave action and hurricane winds. The mangrove

Red mangroves rest on prop roots which support the tree above water as if it were on stilts. Everglades National Park, Florida.

forests that cover the tidal flats and brackish river mouths of southern Florida create 500 square miles of inaccessible wilderness where many rare water birds have their rookeries, and manatees, alligators, and crocodiles find sanctuary. This strange forest also shelters such old friends as raccoons, deer, and bobcats.

The *hammock forests* of this biome are another unusual type of American community. Each hammock is a dense growth of angiosperm trees, either palms or evergreen dicots, which forms an island amid the marshes and swamps. A hammock only a few acres in extent may consist of several dozen different kinds of trees and twice as many shrubs; it is also a luxuriant garden of small flowering plants, ferns, mosses, and epiphytes. There are two kinds of hammocks: one is found at the northern border of the biome, the other in the tropical southern portion. In the northern hammocks are many trees we already know. Live oaks are the dominant trees of the hammocks in higher locations. At intermediate elevations the hammocks are a mixed live oak–cabbage palm forest. The hammocks found on low swampy ground consist entirely of cabbage palms. The trees of a southern hammock community are quite different: they are usually a climax forest of well-established tropical species. Among them we find trees with such intriguing names as strangler fig, poisonwood, and manchineel. Beneath the tree canopy hangs a network of huge vines: rattan, possum grape, pepper vine, rubber vine. Every available branch provides a niche for epiphytic ferns, air plants, and orchids. Hammocks of this type form the most tropical community found in the United States.

It is a crowded community, every available niche being filled by plants adapted for this particular way of life. It is a jungle where survival is not so much a matter of adaptation to the physical environment as of ability to compete with other plants for a roothold and for sunlight. Unusual features are the dominance of broad-leaved evergreens and the preponderance of monocots. Also surprising is the great number of cacti, popularly considered plants only of the desert.

Broad-leaved Evergreens. The landscape of southern Florida is a green one, summer and winter, owing to the great number of broad-leaved evergreen trees, most of them with thick, leathery, and glossy leaves. They occur in all types of communities. Some are the dominant trees in the hammocks, some thrive on sandy coastal sites, and others live on tidal mudflats. In each of these environments we find trees with unusual adaptations suited to special niches in the community.

Many of the trees found both in the hammocks and on the open sandy coastal areas belong to the pea family. SWEET ACACIA is a small spreading tree with spiny twigs, large compound leaves, small yellow flowers in compact spherical clusters. Like all members of the family, the fruit is a pod; that of the sweet acacia is cylindrical and purplish brown. CAT'S-CLAW, as the name implies, is also a spiny tree; it has a similar low-spreading crown, with foliage of compound leaves. The coiled pods contain black and red, shiny seeds. Cat's-claw forms dense thorny thickets on sand dunes close to the saltwater; it is the dominant shrub on the more northern keys. Overtopping the shrubby vegetation along the Overseas Highway is the WILD TAMARIND, another low-growing tree of the pea family. It has thin flat pods reaching a length of five inches.

Some of the trees provide juicy fruits that are eaten by the birds. Common on dunes and beaches is the SEA GRAPE, the fruit of which resembles small purple grapes. It is a tall shrub or small tree with twisted and spreading branches, and red-veined, almost circular leaves. The related DOVE PLUM, with dark red fruit, becomes a larger tree; it occurs in greatest abundance on the keys. The COCO PLUM, a tropical member of the rose family, is another small tree, with purplish plumlike fruit an inch or more in diameter.

Two of the tropical trees have very hard wood: lignum vitae and West Indies mahogany. LIGNUM VITAE is a small tree with a stout gnarled trunk, white bark, and compound leaves. It is one of the few native trees with blue flowers; the fruit is a conspicuous orange-colored capsule. Wood of lignum vitae is so dense that even when dry it sinks in water. WEST INDIES MAHOGANY is a larger tree, often forty feet tall, with scaly brown bark, compound leaves, and small greenish flowers. Most of the large mahogany trees have been cut for lumber, but small trees can still be found on the keys.

Several of the trees in this biome have the unique characteristic of being poisonous. POISONWOOD, an oversized relative of the common poison ivy, grows to be a medium-sized tree; it has compound leaves, small yellowish flowers, and orange-colored berries. The sap of this tree, like that of poison ivy, is a powerful skin irritant. The FLORIDA FISH-POISON TREE is of the same size as poisonwood, and also has compound leaves; but the blossoms are conspicuous pale lavender, pea-shaped flowers similar to wistaria. The powdered bark of this tree was used by the Indians for stupefying fish. The most poisonous plant found in North America is MANCHINEEL, a small tree with a deceptively at-

220 / THE BIOMES OF NORTH AMERICA

tractive apple-like fruit. The sap, highly toxic even when diluted with rainwater dripping from the foliage, can cause infection and blindness; eating the fruit can be fatal. The poison was used by the Indians in making lethal tips to their arrows. When the first white men came to Florida, the Indians are said to have scattered twigs of manchineel in the springs, and placed the tempting fruits on the ground nearby so those who drank the water or ate the fruit would die. A University of Florida biochemist studying the poisonous chemical of this tree accidentally got some of the sap on his hand. The infection that resulted covered his arm with ulcers and caused a partial paralysis.

Strangest of all the trees in this biome are the mangroves, which encircle the southern tip of Florida. The RED MANGROVE is a pioneer; it takes a position on the seaward front of the forest, rooted in mud covered by water even at low tide. Red mangroves have arching prop roots that support the tree above the water as if it were on stilts. Since the

Black mangroves develop slender upright aerial roots that rise from the mud like asparagus tips. Everglades National Park, Florida.

trees grow close together, these prop roots become tangled nets that trap debris, and as a result are constantly extending the shoreline out into the sea. The leathery, evergreen leaves form a dense flat-topped crown; and amid the foliage are borne inconspicuous yellowish-green flowers. The red mangrove reveals an unusual reproductive adaptation enabling the seedlings to survive on the flooded mudflats. The seeds germinate in the fruit, while it is still attached to the parent tree — a viviparous habit comparable to that of a mammal in the animal world. One can see many fruits with fingerlike seedlings, often twelve inches in length, hanging from them; when the seedlings drop into the water, they float in a perpendicular position. As the seedling sinks into the mud, its tip therefore becomes firmly anchored and soon takes root. The BLACK MANGROVE forms a zone behind that of the red mangrove; it grows on mudflats exposed at low tide but covered with water when the tide is high. The black mangrove is usually a larger tree, and has leathery yellowish-green leaves. It has no prop roots, but develops many slender upright aerating roots, known as pneumatophores, which are similar to miniature cypress knees. They cover the muddy soil around the base of the black mangroves like asparagus tips in a vegetable garden.

Palms are a characteristic feature of southern Florida; in some areas they are scattered among the broad-leaved trees, in others they occur among conifers. The trunk of a palm is tall and slender, topped by a cluster of a few huge compound leaves often reaching a length of more than fifteen feet. The flower clusters are also designed on a mammoth scale, hanging in large masses beneath the leafy crown. Since palm trunks have no cambium, or growth tissue, they cannot increase in diameter as do pines and oaks. Palms can be arranged in two groups on the basis of their leaves. In one group, the fanleaf palms, the leaf has a rounded outline and consists of segments radiating in all directions from the end of the leafstalk. In another group, appropriately known as the featherleaf palms, the leaf is long and narrow, and the parallel segments grow from the leafstalk like the barbs on a feather.

Most widespread of the fanleaf palms is the CABBAGE PALM, which we first encountered in the coastal plain forest of the Carolinas. It can stand lower temperatures than other palms and consequently can grow farther north. Although occurring throughout the southeastern United States, it attains its most luxuriant growth in the pine forests, cypress swamps, and Everglades of Florida. Cabbage palms grow to a height of eighty feet, yet have a trunk diameter of only eighteen inches. The

Cabbage palm hammock near St. Augustine, Florida.

trunks of young trees are rough, covered by the broken shafts of old leaves. Each leaf is four to five feet in diameter, borne on a long, smooth leafstalk. The large terminal bud, which resembles a cabbage, is considered a delicacy by native Floridians. A dwarf relative of the cabbage palm, the SAW PALMETTO, a common shrub in sandy pine forests and in the dune-beach community, has grayish green leaves and spiny leafstalks. The trunk of a saw palmetto sprawls over the ground, its upturned tip supporting a cluster of stiff leaves.

The featherleaf palms are often planted as ornamental trees in southern Florida. Of these, the ROYAL PALM is truly a regal tree. With a height of one hundred feet, it is one of the tallest monocots of North America. The smooth gray trunk, two feet in diameter, is swollen at the base; the upper portion is green where the leaf sheaths support the canopy of huge leaves. A particularly fine royal palm grove remains at Collier-Seminole State Park, a hammock at the western end of the

Royal palms at Collier-Seminole State Park, Florida.

224 / THE BIOMES OF NORTH AMERICA

Tamiami Trail. The COCONUT PALM also has smooth gray bark and grows almost as tall as the royal palm. Its leaning trunk and crown of ragged windswept leaves add a tropical background to many of the beaches and keys. Coconut palms are not native members of the biome but, being very adaptable to the southern Florida environment, they have become as abundant as the royal palm.

If we were to choose a single feature that gives this biome its most exotic aspect, it would be the abundance of *epiphytes,* which cover every available limb and tree trunk. High humidity and high temperatures combined with many shaded habitats provide an ideal environment for plants adapted to this niche. Many of the epiphytes are mosses and ferns, but a number of flowering plants have also adopted this aerial mode of life. These belong chiefly to the pineapple and orchid families. In the pineapple family is the ever-present SPANISH MOSS, the leafless

Air plants perch on the branches and in the crotches of trees, the vase-shaped cluster of succulent leaves acting as a trap for rain water.

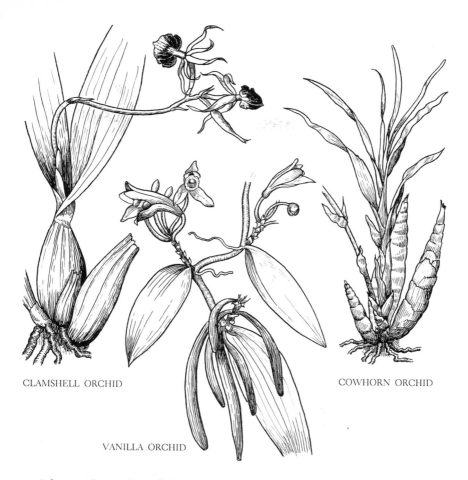

CLAMSHELL ORCHID

COWHORN ORCHID

VANILLA ORCHID

epiphyte whose threadlike stems drape the trees with gray festoons. Other species in the family are known as AIR PLANTS, or wild pines. These have stiff, succulent, pointed leaves that grow in a vase-shaped cluster several feet high. Air plants perch on the branches and in the crotches of trees; the spaces at the base of the leaves act as a trap for rainwater and debris, which aid in nourishing the plant. From the center of the leaf cluster rises a flowering stalk with many small red, purplish, or white blossoms. These add a bright touch to the swamps in early spring, when the fresh green foliage appears on the cypress trees.

The orchid family, with its showy flowers highly specialized for insect pollination, is the elite of the plant kingdom. In the northern states orchids grow in soil and have conventional root systems, but in the tropical biome we find many epiphytic species that have spread to the peninsula of Florida from the West Indies and Central and South America. Their aerial roots obtain necessary nutrients from the air and also provide a firm attachment to the host tree. Most of the epiphytic orchids have stiff or leathery, lilylike leaves; they also develop special food-storage organs known as pseudobulbs. In many species the roots cling, snake-like, to tree trunks and branches. More than twenty species of epiphytic orchids, with flowers in a great variety of colors and shapes, grow in the hammocks and cypress swamps of southern Florida. The beauty of orchid blossoms has led to the near-extinction of many species by irresponsible collectors. As a result, some of the orchids have become very rare.

Most widespread is the GREENFLY ORCHID, commonly found on oak and cypress. It has small brownish-green flowers borne in a compact cluster. One of the most peculiar species is the COWHORN ORCHID, whose large pointed pseudobulbs are responsible for its name. The yellow flowers, spotted with brown and purple, form a very large loose spray, which looks much like a swarm of insects. Cowhorn orchid is common throughout the cypress swamps and hammock forests. In pinelands and hammocks we find the SPREAD-EAGLE ORCHID, the flowering stalk grow-

GREENFLY ORCHID

WHITE BUTTERFLY ORCHID

ing to a length of several feet and bearing a spray of yellowish flowers blotched with magenta and brown. The spread-eagle orchid develops only one large leaf. Most rare and elusive member of the family is the dainty WHITE BUTTERFLY ORCHID, found on live oaks and royal palms. It is unusual in being entirely leafless; photosynthesis is carried on by the flattened green roots, which cling closely to the tree trunk and so perform a dual function. One orchid enthusiast has described this spirit of the tropical hammock by saying, "When one looks into the gloom of a thickly leaved tree and sees the extraordinary flowers of this little orchid for the first time, one is instantly impressed with its likeness to a thin flat snow-white frog suspended in mid-air — caught, as it were, in the middle of a leap from one branch to another." Most vine-like of all the orchid epiphytes is the ZIG-ZAG VINE, or leafy vanilla vine, common in the Big Cypress Swamp. It has special aerial roots that cling to tree trunks, and flat elliptical leaves. Each greenish-gray flower has a conspicuous flaring lip. A related species, the VANILLA VINE, the pods of which supply the well-known flavoring extract, is sometimes found as an escape from gardens.

One of the strangest adaptations for jungle survival is revealed by the STRANGLER FIG, which begins its life as an epiphyte. The seed germinates in the upper branches of a tree,

The strangler fig eventually kills its host by means of the encircling roots and canopy of foliage.

usually a cabbage palm. Appropriately called "the strangler," the seedling quickly develops aerial roots that grow downward around the trunk, encircling the tree in a fatal embrace. Reaching the ground, the seedling takes root and soon produces a canopy of foliage which overshadows the host, and eventually kills it by shutting out the light. Such reproduction is a remarkable adaptation in a competitive environment where finding a "place in the sun," a prime requisite for survival, is a life-and-death struggle for a seedling tree. Strangler figs are related to the tropical banyan tree, whose spreading branches send down aerial roots that in time become additional trunks. Thus a single tree can become a small forest.

The cacti of southern Florida do not dominate the landscape as they do in the desert biome, yet they occur in sufficient number and variety to add another unusual aspect to the biome. Most widespread are the PRICKLY PEAR CACTI, of which more than thirty species occur in the southeastern states, many of them restricted to southern Florida. They are primarily xerophytes of dry habitats, where they spread their spiny pads over sand dunes or among grasses and saw palmetto. In more mesophytic situations they become shrubs and small trees. Most prickly pear cacti have showy yellow flowers and produce juicy edible fruits resembling small pears, eaten by both Indians and animals.

Other species of cacti have cylindrical fluted stems whose lengthwise ridges are armed with a formidable array of sharply pointed spines. Many are weak-stemmed plants that clamber like vines over shrubs or climb the branches of trees; in most species the flowers are large and showy, blooming at night. In coastal hammocks on the keys grows the BARBED-WIRE CACTUS, its trailing stem as slender as one's finger yet growing to a length of twenty feet or more. It has angular edges, with prominent lengthwise ridges; often the tip arches over and takes root, making a thorny thicket through which progress is impossible. The MISTLETOE CACTUS, at home in the hammocks of the Florida Keys, clings to trees by means of a tangle of cylindrical or flattened stems; its berries resemble those of mistletoe.

Two unusual cacti of this tropical region are the tree cactus and the lemon vine. The TREE CACTUS has an erect, unbranched, and fluted trunk that rises to a height of thirty feet. Tree cacti grow in rocky hammocks on the keys; all have near relatives in the West Indies and tropical America. Another strange cactus is the LEMON VINE, also found on hammocks of the keys. Although a cactus, the lemon vine bears glossy elliptical leaves like those of any typical dicot; beneath each leaf a cluster of cactus-like spines grows from the stem. This leaf-bearing species represents the primitive cactus as it was before leaflessness became a characteristic adaptation for life under arid conditions.

The Living Environment: The Herbivores

The luxuriant growth of vegetation produces more than sufficient food for the herbivores of the ecosystem. Among these primary consumers are many invertebrates: worms, plant-feeding insects of both land and water, and mollusks. The omnivorous songbirds from the North, which winter here, eat fruits, berries, and the seeds of palms and dicot trees. That prize game bird of the woodlands, the wild turkey, is also one of the herbivores of the Florida forests. Among the mammal herbivores are the rodents and rabbits, which are common throughout this biome, as they were in the coastal plain forest. In the flatwoods and hammocks we can find such old friends as gray squirrels, harvest mice, cottontail rabbits, and white-tailed deer. Three additional, large southern rodents live in the marshes: the rice rats, cotton rats, and Florida water rats. The grayish or light brown RICE RAT reaches a length of twelve inches. Slightly larger is the dark brown COTTON RAT, a pest in the sugarcane fields surrounding Lake Okeechobee. The many swamps would seem an ideal habitat for muskrats, but this niche is pre-empted by the FLORIDA WATER RAT. It is more of a land dweller than the muskrat; it is also smaller, and has a round rather than flattened tail. The Florida water rat builds a crude nest of rushes, grasses, and bark of mangroves in shallow water or on a floating mat of vegetation near its feeding ground.

The tropical biome is the home of two unique herbivores not found elsewhere on the continent. These are the tiny key deer and the water-dwelling manatee. KEY DEER are a dwarf variety of the white-tailed deer. This small ungulate is about half the size of its northern relative, a mature buck weighing less than fifty pounds. Key deer have the most restricted range of any mammal, being found only on a few of the Florida Keys. Here they roam among mangroves and cabbage palms, with such odd companions as alligators and egrets. The key deer were on the verge of extinction in 1940, when their numbers had decreased to forty individuals. But in 1957 a thousand acres on Big Pine Key were set aside as a National Key Deer Refuge; here these rare herbivores are protected, and have now exceeded 200. Today the greatest threat to the key deer race is sudden death from the ceaseless automobile traffic that speeds across the key on the Overseas Highway.

Undoubtedly the most bizarre of all our native mammals is the MANATEE, or sea cow. This strange creature belongs to a small order of aquatic mammals known as sirens, which also includes the extinct Steller sea cow of arctic waters. In North America the manatee is a vanishing race, being found chiefly in a few lagoons and shallow bays off the coast of southern Florida. Much of its habitat is fortunately within the boundaries of Everglades National Park. Manatees are huge animals, reaching a length of thirteen feet and weighing as much as half a ton. The plump rounded body of the manatee is dark gray, sparsely covered with bristly hairs; bristles also surround the broad mouth with its prehensile and cleft upper lip. The eyes are small and there are no external ears. The manatee's ancestors forsook the land to live in the sea, and acquired many adaptations for an aquatic mode of existence. The forelimbs have become paddlelike flippers; the hind limbs have entirely disappeared. A flattened tail, terminating in horizontal flukes, has become the main organ of locomotion. The manatee swims by vertical sweeps of its tail, browsing lazily on the abundance of aquatic plants found in shallow water. Its life is a leisurely one spent in lolling in the warm water in a vertical position, moving only when in need of another meal of grass. Manatees must come to the surface to breathe, but can hold their breath underwater for ten to fifteen minutes at a time, even as long as thirty minutes. The greatest hazard to their existence is cold weather. Not having a protective coat of blubber, as most other sea mammals do, manatees cannot stand freezing temperatures. Their number is reduced also by predatory sharks and crocodiles, which feed on the young.

THE LIVING ENVIRONMENT: THE CARNIVORES

Because much of this biome consists of marshes and swamps, the carnivores of the ecosystem include a great number that live in or near the water. Feeding mainly on the invertebrate herbivores is a great variety of amphibians, reptiles, and water birds. Many are the same as those of the coastal plain to the north, but with them we find tropical species that have made themselves at home in this southeastern tip of the United States. Mammal carnivores are few and are inconspicuous members of the wildlife community. Although bobcats, black bears, and mountain lions live in the deep woods, they are rarely seen.

Some of the carnivorous reptiles we have already seen in the deciduous biome. Among these are the scurrying fence lizards and colorful anoles sunning themselves on the trunks of the trees, pond and mud

The tiny but extremely venomous coral snake is an inhabitant of the edge of the tropics, in Florida as well as in the southern coastal plain forest of adjacent states. (*Credit: Courtesy of The American Museum of Natural History*)

turtles on the banks of the numerous roadside waterways. Cottonmouths, or water moccasins, coil about the cypress knees; in the pine-oak woods, diamondback rattlesnakes hide in the prickly saw-palmetto brush; and coral snakes lead a sequestered existence in the litter of the hammock floor. Two harmless snakes are especially worthy of note. The SCARLET KINGSNAKE is outstanding as an example of the extraordinary mimicry that occasionally exists in nature: the coloration and general appearance of the kingsnake simulates that of the poisonous coral snake. Both have crossbands of red, black, and yellow. In the scarlet kingsnake the narrow yellow rings are bordered on each side by black bands; in the coral snake, however, each yellow ring lies between a black and a red band. The harmless kingsnake lives in pine woods, usually near water, where it finds worms, small fishes, lizards, and mice. The INDIGO SNAKE is remarkable for its size and amiability; it is one of the largest and yet the most gentle of all our American snakes. This sleek blue-black species grows to a length of more than eight feet, an impressive size for any animal. Indigo snakes are most abundant in the Everglades region, where they feed on lizards, rodents, and other snakes. Their docile disposition is contributing to their extinction; many indigo snakes are collected for sale as pets and for use by snake charmers.

Of all reptiles, the most powerful and dangerous are the crocodilians, native to subtropical and tropical regions throughout the world. They are heavy-bodied animals with stubby limbs, their bodies protected by a bulletproof armor of leathery plates and their muscular tail strong enough to knock a man off his feet. Their elongated head ends in a long snout with huge mouth and powerful jaws armed with a fearsome array of sharp teeth. Crocodilians are well adapted to reign as top aquatic carnivores. They swim effortlessly by sidewise sweeps of the long tail, and can float near the surface, completely submerged except for the projecting snorkel-like nostrils and the periscope eyes. Lying thus, a crocodilian remains undetected; when an unsuspecting animal comes in sight, the reptile sinks slowly below the surface, swims beneath the prey, and captures it with a quick sidewise movement of the powerful jaws. A crocodilian is able to swallow its prey underwater, because of a special valve that can close the nostrils, and a flap of skin which can shut off the mouth cavity from the windpipe.

In the United States the crocodilians are represented by two species, the alligator and the crocodile. The alligator is a temperate-climate reptile that adapts to low temperatures by becoming dormant like a hibernating mammal; it reaches its *southern* limit in Florida. The crocodile, on the other hand, is very sensitive to cold; water below 45° F. is usually fatal. It is a tropical reptile reaching its *northern* limit in Florida. And so the extreme ranges of these two crocodilians overlap in this biome. The AMERICAN ALLIGATOR is a dark brown or black reptile with a broad flat snout, found in cypress swamps, ponds, and streams. Alligators may reach a length of nineteen feet and a weight of almost 1000 pounds. They eat anything provided by nature or man: their normal diet consists of fishes, turtles, water snakes, herons, and egrets. Formerly common from North Carolina to Texas, alligators were almost exterminated by hide-collectors and "sport" hunters; now they are protected in Georgia and Florida. The AMERICAN CROCODILE, which has a saltwater rather than freshwater habitat, differs from the alligator in having a much more pointed snout. Crocodiles are smaller than alligators, usually being under twelve feet in length, but are more belligerent and dangerous. It is fortunate that these two unusual reptiles have found sanctuary within the boundaries of Everglades National Park.

Our drive along the Tamiami Trail leaves the impression that the roadside marshes are one vast zoological park patrolled by thousands of long-legged wading birds. They are present in such great numbers because of the abundance of animals on which these feathered carnivores subsist. The marsh water teems with crustaceans and mollusks, fishes and frogs, water snakes and turtles — a feast for birds adapted to live in

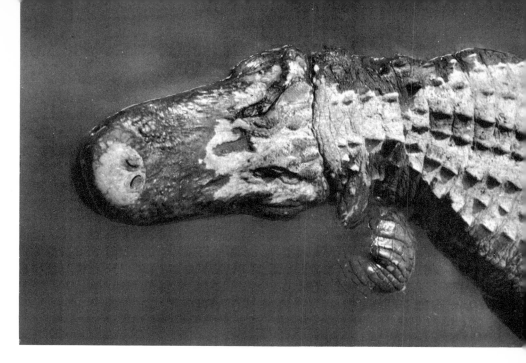

The alligator (upper) has a blunt snout; the crocodile (lower) has a tapering head with a pointed snout.

ANHINGA

this environment. The coastal beaches and mudflats, alive with crabs, crayfish, worms, and snails, provide a repast for the shorebirds. The dominant warm-blooded carnivores are the birds. For them the Everglades, the swamp forests, and the hammocks are an avian paradise.

The domain of the water birds is explored to best advantage by boat. In a small craft, surrounded on all sides by water weeds, grasses, cypresses, and mangroves, we can thread our way through streams and channels that form a labyrinth of waterways. Picturesque cabbage palms lean over the water, clumps of palmetto edge the banks. Above our heads a tangle of branches is weighed down with air plants and ferns. We move through a tunnel of green, and everywhere are long-legged birds — anhingas, limpkins, gallinules, egrets, ibises.

Our most constant companion is the ANHINGA, also known as the water turkey or snakebird. This large black bird has a long flexible neck, and head terminated by a sharp spearlike bill. We pass a motionless anhinga perched on the lowermost branch of a tree, its wings spread to their full width of four feet to dry in the sun. Another anhinga is

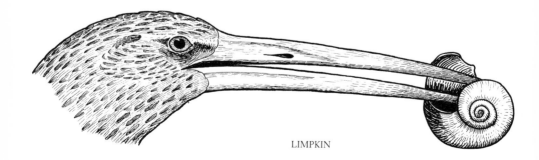

LIMPKIN

alertly scanning the water beneath its perch; as a fish comes into view, the bird plunges downward with lightning speed and reappears with its prey grasped firmly in its bill. Other anhingas swim in the water, most of the body submerged. The bird's long neck and head held above the surface bear a startling resemblance to a large black snake. Continuing on our way, we see a crane-like bird moving awkwardly with a limping gait through the riverside grasses. It is a LIMPKIN searching for snails. At our approach the bird rises clumsily into the air, legs dangling. It barely clears the marsh grass and soon drops out of sight in the swamp vegetation. Limpkins are most active at night, when their weird wailing cry explains the local nickname of "crying bird." Around a bend in the stream, we come upon a quiet cove, the surface covered with lily pads. Walking as confidently on this floating mat as if it were solid land, is the chicken-sized, dark blue and green PURPLE GALLINULE. This "gem of the marshlands" can move on such insecure footing because of long wide-spreading toes. It is a fearless bird, allowing us to approach so closely that we can admire its colorful combination of yellow legs, white cap, and red patch at the base of the bill. Among the floating vegetation, gallinules find an almost unlimited supply of insects, snails, frogs, and lizards.

At every turn we glimpse white wading birds. On a shallow sandbar we surprise a COMMON EGRET watching the water intently, daggerlike bill poised in readiness to capture an unwary fish or frog. This snow-white bird stands three feet high, its yellow bill and black legs a contrast to the spotless plumage. When it takes to the air, its wings reach a span of more than four feet. During the breeding season, egrets develop long filmy plumes known as aigrettes, and once highly prized as decoration for women's hats. Millions of these birds were killed by ruthless plume-hunters who supplied the millinery market with the aigrettes. The species would undoubtedly have become extinct in the United States were it not for the efforts of the National Audubon Society to secure passage of laws protecting the egret and outlawing the aigrette industry. It was not an easy victory. At Florida's Cape Sable is the grave of an Audubon warden, Guy Bradley, murdered by a plume-hunter in 1905 while attempting to enforce the law. Wherever conservation-minded men attempt to stop senseless destruction of our plant or animal neighbors, they find it a long struggle against the interests of those who value our wildlife only in terms of dollars and cents.

Another white wading bird that we find along our course is the SNOWY EGRET, a smaller species standing only two feet high. It is dis-

Common egret. (*Credit: A. D. Cruickshank, from National Audubon Society*)

tinguished by its black bill: the legs are also black, but the feet are yellow. This is a nervously active bird, continually on the move, stirring up the water with its feet to uncover shrimp, crayfish, and crabs. This egret also develops special plumes during the breeding season, and so it too came perilously close to extinction before the aigrette traffic was prohibited. As we come to the end of our boat trip through the Everglades region, we see a huge bird perched in the topmost branches of a cypress. This is the WOOD IBIS, the only American member of the stork family. It is an impressive bird, standing four feet tall and having a wingspread of five and one-half feet. The black head and neck are bare and scaly; because of this, the birds are called "flintheads." The body is white, but the wings have black edges, tail and legs are also black. The wood ibis prefers habitats where swamps alternate with shallow ponds and mudflats; here it finds ample food in the form of crustaceans, amphibians, and small reptiles.

One of the two largest wading birds in North America is the GREAT WHITE HERON. We find this bird on the small mangrove islands off the southern tip of the peninsula and on the keys. It stands over four feet in height and has a wingspread of six feet; it differs from the other white wading birds in having both bill and legs yellow. During the breeding season, its head is decorated with long crest plumes. The white heron, sedate and slow-moving, feeds on shrimp and other crustaceans, as well as on fishes found in muddy bays and on tidal flats. They nest in sites difficult to reach, but even there are not safe from native fishermen, who consider the nestlings a delicacy. The disastrous hurricane of 1935 reduced the white heron population to less than 200 birds. A few years later the Great White Heron National Wildlife Refuge was established, and today the population has increased to several thousand.

Here at the edge of the tropics live several species of the unusual ibis family, whose members can be recognized by their downcurved bill. The WHITE IBIS is a regal bird with a touch of black on the tips of the wing feathers; its bill and legs are red, and the bare flesh around the base of the bill adds another patch of red. The size of a small egret, the white ibis is only two feet tall. Usually seen on mudflats, it hunts for crabs and crayfish when the tide is low. The white ibis is very sensitive to invasion of its home grounds by man: a rookery may be abandoned after only a single cautious visit by a human. Its rookeries therefore should be carefully protected. Of all the ibis family, the ROSEATE SPOONBILL is the most rare and the most beautiful. It is attired in pastel

The flamingo was formerly a common Florida visitor from the Bahamas and Cuba, but never nested in the United States.

ROSEATE SPOONBILL   WOOD IBIS

shades of pink and white, with greenish-black head; the legs are red and the bill is green. The unusual bill is flattened and is wider at the tip than at the base; it appears clumsy compared with herons' bills, but is put to efficient use by the roseate spoonbill. Feeding in shallow water, the ibis shovels its way through the mud by sidewise movements of the wide bill. By opening and closing this bill, the spoonbill strains small aquatic animals out of the water. The spoonbill is now practically extinct in Florida, but a small population survives in a wildlife refuge in Texas.

The visitor to southern Florida cannot fail to be fascinated by the AMERICAN FLAMINGO, an attraction at many of the winter resorts. This bird was formerly a common visitor from the Bahamas and Cuba, but never nested in Florida. A flamingo stands four feet tall, and has a five-foot wingspread; its plumage is a rosy pink with black margins to the wings. The bill, awkward-looking with a bent tip, seems misshapen until we discover that the flamingo feeds with its head upside down. Then we can see that the strange shape is a useful adaptation, the bent por-

BROWN PELICAN

tion serving as a scoop to sweep in aquatic prey, chiefly mollusks, which are eaten whole.

Another large water bird of the biome, common along the beaches, is the BROWN PELICAN. This ungainly bird has the tremendous wingspan of six and one-half feet. Head and neck are colorfully striped with brown, yellow, and white; the rest of the plumage is brown. In flight, the pelican hunches head and neck into its shoulders. Alert for fishes, squadrons of pelicans patrol the shallow water at the edge of the beach, soaring close to the tops of the waves and dipping into the troughs. When a fish is seen, the pelican plunges for it with a noisy splash. The overdeveloped sac beneath the bird's lower jaw serves as both a scoop and a place to store food. Brown pelicans have acquired the role of scavenger, and rare is the pier that does not have at least one of these sentinels on lookout, ready for an easy meal.

All these water birds are carnivores depending upon invertebrates and aquatic vertebrates for food. Thus the flow of energy passes from the crustaceans, mollusks, fishes, frogs, and water snakes, into the herons, egrets, ibises, and pelicans. Another type of food chain leads from toads, snakes, lizards, and rodents to the birds of prey. This terminates in such feathered carnivores as the owls, hawks, bald eagle, and vultures. We have already met many of these in the deciduous forest biome. The hawk family, however, includes two members we have not encountered elsewhere: the swallow-tailed kite and the Everglade kite. The SWALLOW-TAILED KITE is a large hawk with a four-foot wingspread, easily recognized by its long forked tail. It is a conspicuously black and white bird; the head and underparts are white, the back, underwing edges, and tail are black. This kite is a bird of river-bottom swamps and forests, catching most of its prey in flight. It is an accomplished soaring bird but can also swoop swiftly to tree level and snatch a lizard or treefrog deftly from the branches. The swallow-tailed kite was formerly widespread in the southeastern United States but today is rare except in Florida. The EVERGLADE KITE is a dark brown hawk with square tail and broad wings; it is smaller than the swallow-tailed kite. In spite of its name, the Everglade kite is no longer found in the Everglades. In earlier years it was abundant, but the building of drainage canals altered the habitat so that its staple food, a large land snail, disappeared. The few now remaining live near Lake Okeechobee. Flapping slowly over the tops of the marsh grasses in this region, the Everglade kite picks off snails and carries them to its perch in a tree; there, with long hooked bill, the kite extracts the meat without breaking the snail shell.

Man and the Tropical Biome

Fortunately for the wildlife of this biome, the extensive marshes and swamp forests, the generally poor soils, and the lack of marketable timber have discouraged exploitation. Much of the southern tip of Florida is still a wilderness, and the creation of the Everglades National Park guarantees that the large area within its boundaries will continue to be a sanctuary for wildlife. Passage of laws to protect the plume-bearing herons, the alligators, manatees, spoonbills, and other rare creatures that would otherwise soon be extinct has also assured their continuing survival — provided their habitats remain unchanged. Present attempts to drain large tracts of swamplands, in order to gain more land suitable for use by man, lowers the water level of the Everglades basin and becomes a threat to the wildlife living in the marshes and swamps of this area. If such drainage continues, it may be fatal to much of the southern Florida wildlife.

Conflict of interests between man and the wildlife community is inevitable. In every biome it brings special problems. In the conifer forests, the conflict involves the destruction of redwoods and pines for lumber, the modification of the environment by highways and recreational facilities. In the deciduous forests, the problem has been the growth of cities and the resulting need for more living space and increased agricultural acreage. Here in the tropical biome it is the control of the waterflow through the Okeechobee-Everglades basin. When we explore the grassland and desert biomes, we shall find there still another conflict of interests. In all cases the fundamental issue is the same: how best to serve all needs by modifications of the habitat in harmony with the long-range interests of the entire ecosystem — of which man is only a part.

12 / The Grasslands

THE DECIDUOUS FORESTS of eastern North America are separated from the conifer forests of the West by a broad corridor of grassland several thousand miles in length and more than 500 miles in width. When the first pioneers left the shelter of the oak-hickory woodlands, they found themselves at the edge of this vast expanse of unexplored land. The grassy plains that extended to the far horizon seemed an endless green sea, rising and falling like the swells of the ocean. The eastern portion of this grassland biome, where the grasses grew taller than a man, is known as the *true prairie*. It forms a north-south strip adjacent to the deciduous forest, and includes Illinois, Iowa, Missouri, and parts of Wisconsin and Minnesota. The original prairie has now disappeared to become our leading grain-raising area. Farther west, at the 20-inch isohyet and approximately at the 1500-foot-elevation contour, lies the *mixed prairie*. Here the dominant grasses are medium-height species restricted to low moist sites and shortgrasses that grow on the drier uplands. On the semi-arid Great Plains, reaching to the base of the Rocky Mountains, the characteristic community is the *shortgrass plains*. Included in the mixed prairie and shortgrass plains are the western portion of the Dakotas, Nebraska, and Kansas, and the eastern part of Montana, Wyoming, and Colorado. Suitable for grazing rather than farming, this region has become the center of our cattle empire. To the southwest, a *desert grassland* occupies the more arid plains of western Oklahoma and Texas, parts of Arizona and New Mexico. This is also a shortgrass community, bordering the desert, from which such xerophytes as cactus, yucca, and mesquite have spread into the grassland biome.

The origin of the grasslands dates back at least twenty-five million years, to the period when geologic forces were building up the Rocky Mountains. This uplift changed the climate of the interior of the continent, reducing precipitation and creating unstable weather conditions with unpredictable periods of severe drought. As the environment became unfavorable to tree growth, the original forests disappeared. Herbaceous perennials, especially grasses, became the dominant vegetation

THE GRASSLAND BIOME

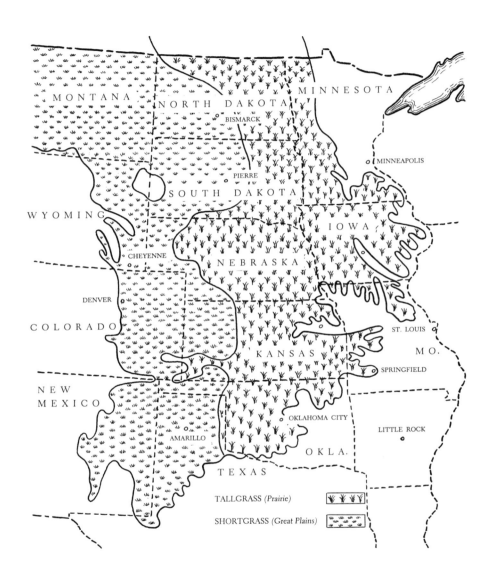

since they possessed the adaptations essential for survival in the new climate. These vast grasslands provided food for the largest assemblage of herbivorous mammals the earth has ever seen. The prehistoric American prairie was the home of camels and elephants; here the entire evolution of the horse took place, from its diminutive four-toed ancestors. Later the colder climate of the glacial periods brought extinction to these primitive mammals; it also enabled the boreal forest to advance southward into the high elevations of the biome, where their descendants remain today as scattered stands of white spruce, paper birch, and aspen poplar.

The Physical Environment

To the casual observer the grassland biome may seem an uninteresting expanse of land covered with nothing but grass. It is, however, an intricately organized community wherein the climax vegetation — the grasses — has become completely adapted to the physical factors of the environment. These factors, as in the other biomes, include precipitation, temperature, topography, and soils. The wildlife community that exists today has resulted from a slow sorting out of species over millions of years, with survival of those best fitted to flourish on the interior plains of the continent.

The most critical physical factor is precipitation. In the grassland the competition among plants is for water, not — as in the deciduous biome — for light. At the eastern edge of the grassland, where the true prairie borders on the deciduous forest, the precipitation is forty inches a year. It gradually decreases westward and finally becomes only ten inches in the rain shadow of the Rocky Mountains. The true prairie, in spite of being in a region of ample rainfall, is treeless because of the high winds and excessive transpiration, frequent droughts, and constant fires, which are fatal to tree seedlings yet harmless to grasses. An additional barrier to tree invasion is the ability of the sod-forming grasses to monopolize the ground so completely that tree seeds find it difficult to germinate. At the western edge of the biome, where the precipitation is very scanty, xerophytic trees such as mesquite and juniper are able to establish themselves especially where the native grasses are removed. Because of the wide range of rainfall throughout the biome, different kinds of grasses make up the vegetation; some are best adapted to grow in the moisture-laden soils of the lowlands and river valleys, others to thrive on dry gravelly ridges and hillsides.

The grassland biome extends through three temperature belts. In

Alberta, Saskatchewan, and the north-central United States the winters are long and cold, the summers short and cool. In the central prairie states the winters are also cold, but the summers are hot. In the Texas-Oklahoma region the winters are milder, the summers even hotter. Yet grasses have been able to adjust even to such extremes of winter and summer temperatures. Over the northern part of the grasslands snowfall is often deep, blizzards are common, and the soil freezes to a great depth. This presents a problem in survival for both the plant and animal members of the community. During cold weather and periods of drought, the prairie retreats underground and the landscape is devoid of most of its plant life. But with the coming of warm weather and the rainy season, the grasslands become transformed quickly into a living green world. All the grass species do not respond to these environmental changes in the same way and at the same time. Those of northern origin such as needlegrass or Junegrass resume their growth early in spring and mature in early summer; during the remainder of the hot dry summer they become dormant, but come to life again in autumn and stay green in spite of frost. The southern species, of which buffalo grass is an example, wait until later in spring to renew their growth, and remain green all summer. In this way the two types of grasses supplement each other and provide forage for the herbivores during most of the year.

The topography of this biome is chiefly a gently rolling plain with broad valleys in the eastern portion, canyons and ravines in the western part. The soils are deep and fertile in the true prairie but in the Great Plains are made up of coarser sands, gravels, and clays. The sediments washed from the Rocky Mountains form great alluvial fans over which a growth of prairie grasses spreads. As long as this ground cover is undisturbed, the alluvial sediments remain as a plateau gently tilted downward toward the east. But after excessive grazing and plowing, wind and water can erode the surface into gullies and ravines, badlands and sandhills. At its western edge, where the biome reaches an elevation of 6000 feet, it merges with the lower conifer forests.

The Living Environment: The Plant Life

The uniform appearance of the vegetation of the grasslands is evidence of the extent to which plants with a suitable life form — in this case, the herbaceous perennial — can dominate a community. The biome owes its characteristic aspect to the several hundred different species of grasses that represent minor variations on this theme, each having attained an equilibrium with specific habitats of the environment. We notice many other herbaceous perennials mingling with the grasses, some of them wildflowers common also in the open fields and along the dry roadsides of the eastern United States. Small woody perennials too are prominent, some of them shrubby, others growing to be small trees. It is grass, nevertheless, that dominates the scene. Grasses are far less dramatic in appearance or obvious activities than redwoods or epiphytic orchids. Yet they are extremely versatile forms of life, revealing many special adaptations for the prairie way of life.

General Features of Grasses. The grass plant has perfected a type of leaf, a habit of growth, and a form of root system which combine to give it a distinct advantage in the prairie community. The narrow elongated leaf of a grass grows vertically, so that every leaf blade is able to obtain some illumination during the day, without completely overshadowing the neighboring leaves. Because of the shape and this arrangement of the leaves on the stems, a single acre of grass plants presents ten acres of photosynthetic surface, bringing about a maximum energy flow into the grassland ecosystem. Another distinctive feature of grasses is the method by which the leaf blade grows continuously at its base. As an animal eats the terminal portion, basal growth soon restores the leaf to normal size. This explains why grass can be mowed or grazed without harming the plant, provided enough of each leaf remains to carry on food manufacture for the plant. Such a method of growth of foliage and constant rejuvenation makes possible a continual source of food for herbivores.

The visible part of grass is but a small portion of the plant; hidden beneath the ground is an amazing extensive and effective root system. A grass plant rising but a few inches above the ground may have roots five feet deep, and pervading an area for many feet around the plant. Measurements of the root threads in five square feet of bluegrass sod revealed the astonishing total length of 106 *miles*. The roots of lowland grasses often penetrate to a depth of ten feet, reaching permanently moist soil even when the upper few feet of the ground is dry. The roots of upland grasses, such as buffalo grass, are finer and more branched near the surface, forming a fibrous network in the upper few inches of the soil. This type of root system can absorb sudden rains, an ability that promotes rapid spring growth but requires a dormant period in summer when the soil is dry.

Special Adaptations of Grasses. Obtaining as much water as possible, developing drought-resistant foliage, and retiring underground into a dormant state when water is not available — all are important to grasses for survival. Some grasses have hairy leaves, others have leaves that fold up or roll inward; in both cases water loss by transpiration is reduced. Many grasses develop horizontal underground stems known as *rhizomes*; when the aboveground part of the grass plant dies during a hot dry spell, buds on the rhizomes remain alive. As soon as favorable growing conditions reappear, these buds sprout into new foliage.

Grasses are well adapted not only to survive drought and the grazing of herbivores but also to monopolize the ground against all intruders. This they can accomplish by their special methods of propagation, in addition to producing seeds. The rhizomes, which radiate out from the parent plant just beneath the surface, produce new plants, thus forming a compact sod. This is the method by which wheatgrass and witchgrass spread over an area. Many of the weed grasses are difficult to eradicate from lawns and gardens because of a similar use of rhizomes. Another method of pre-empting living space is shown by species that produce stolons. A *stolon* is a horizontal stem spreading out on the surface of the ground, and at intervals it develops buds. These buds grow into new plants, which take root; eventually a sod is formed. Still a third method is developed by the bunchgrasses. These produce *tillers*, which are side branches growing out in a cluster at the base of the plant; the tillers form a clump of shoots which overshadows the surrounding area and discourages the intrusion of other plants.

Vegetation of the True Prairie. The moist lowlands of the true prairie support a lush growth of tall and medium-height grasses. Most conspicuous is the BIG BLUESTEM, its stems covered with a purplish waxy bloom. This warm-climate species, typical of the grasslands of Kansas, has leaf blades twenty-four inches long, so tender that they are eaten by preference before other grasses; so big bluestem is often the first species to disappear as a result of overgrazing. This representative tallgrass grows to a height of ten feet. Almost as tall is SWITCHGRASS, a coarse sod-forming grass that thrives on drier slopes, where the big bluestem cannot survive. KENTUCKY BLUEGRASS, introduced from Europe, has spread widely throughout this part of the biome. It grows vigorously early in the season but becomes dormant during the hot dry weather of midsummer. Lowland habitats support in addition a number of flowering perennials: phlox, vetch, sunflower, aster, and goldenrod among many others.

The drier uplands are overgrown with grasses of medium height. LITTLE BLUESTEM, a warm-climate grass, is a common member of this community. The leaves are smaller and more slender than those of big bluestem, and the plants are only two or three feet tall. It produces tillers in such abundance that a ground cover of compact sod is quickly formed. JUNEGRASS, a cool-climate grass, is abundant in the central and northern portions of the prairie. It has soft corrugated leaves, and grows to the same size as the little bluestem. This species and little bluestem make up more than half of the plants in the true prairie community.

Vegetation of the Mixed Prairie and the Plains. The greatest proportion of the grassland biome lies in the semi-arid and higher Great Plains region where the climax community is one of medium-height and shortgrasses. The roots of both extend to the same depth and thus each type has an equal opportunity to obtain water. The medium grasses form an upper story, the shortgrasses a lower one. Except in the few low moist sites, the grasses do not form a continuous ground cover as they did in the true prairie. As we move westward beyond the 20-inch isohyet, little and big bluestem disappear, and the drought resistant wheatgrass and blue grama grass become dominant.

The medium-height grasses are cool-climate species and hence more common in the northern part of the mixed prairie. NEEDLE-AND-THREAD

Mixed prairie vegetation at Duncan, Oklahoma. (*Credit: Soil Conservation Service, U.S.D.A.*)

Blue grama grass thrives on top of a mesa in New Mexico. (*Credit: Soil Conservation Service, U.S.D.A.*)

is a bunchgrass with narrow ridged leaves, rough on the upper surface. This abundant species, found from Alaska to Texas, grows to a height of three feet. Dormant during hot weather, it resumes growth with the fall rains and stays green all winter. Its unusual name refers to the sharply pointed seeds with long twisted appendages called awns, resembling a threaded needle. The awn absorbs water during wet spells and straightens out; upon drying it becomes twisted, and in changing its shape it drives the seed deep into the ground. WESTERN WHEATGRASS is a sod-forming species that grows to the same height; it produces long rhizomes, by which it colonizes large areas of ground. The stiff bluish-green leaves roll inward during dry spells, an adaptation to prevent water loss by transpiration.

The shortgrasses of this part of the biome are warm-climate species and consequently more abundant in the southern grasslands. BLUE GRAMA is a drought-resistant grass that can grow on sandy soils; its range is from Texas northward. Only five or six inches high, blue grama has an extensive root system and also produces tillers abundantly. Therefore it serves as an excellent soil binder in areas needing protection against erosion. BUFFALO GRASS, also less than six inches tall, has narrow hairy leaves that grow so close to the ground it cannot be eaten entirely by grazing animals. Like blue grama, it becomes dormant during drought. Its tough, wiry roots penetrate deep into the soil, but its surface fibrous roots spread widely around the plant. Stolons, running along the surface of the ground, produce new plants that aid in colonizing the area. Buffalo grass controls erosion of the topsoil by wind; it served an additional purpose for the early settlers, who used it to cover their sod huts.

Although other types of perennials are in a minority, they often form conspicuous parts of the community. More deeply rooted than the grasses, they survive dry periods and can support aboveground woody stems. SILVERLEAF PSORALEA, one of the tumbleweeds, is a persistent invader of the grasslands. Herbaceous perennials include a colorful array of thistles, asters, goldenrods, and poppies. Numerous cacti can compete with the grasses during droughts, when less xerophytic plants must become dormant. These include many kinds of prickly pear cacti, and their shrubby relatives the CANE CACTI and CHOLLAS. Avoided by most herbivores, they can increase in numbers rapidly. Their reproductive habits also enable them to compete with the grasses. The spiny stem segments break off and cling to the hoofs of grazing animals,

Mesquite and cacti have taken over 55,000,000 acres of original prairie in Texas, as a result of excessive grazing. Cotulla, Texas. (*Credit: Soil Conservation Service, U.S.D.A.*)

which carry the fragments to new locations, where they grow into new plants. SOAPWEED, a species of yucca, is a distinctive plant of the arid grasslands, with its basal clusters of stiff and sharply tipped, succulent leaves. Many kinds of SAGEBRUSH, woody perennials that grow to be bushy shrubs, have a gray-green foliage that contributes the dominant color to large areas of the western grasslands. MESQUITE, a stunted tree of the pea family, with compound leaves and pods for fruits, spreads into the grasslands when the grasses that normally cover the ground disappear through excessive grazing. This troublesome invader has taken possession of 55,000,000 acres in Texas alone, since the introduction of cattle into that state.

THE LIVING ENVIRONMENT: THE HERBIVORES

The prairies are a paradise for multitudes of herbivores because of the abundance of grass and of the seeds and fruits of other herbaceous perennials. Among these herbivores are many insects, birds, and mammals. The insects belonging to the grasshopper clan are particularly common; with their specialized scissorlike mouthparts they can cut their way through an astonishing amount of grass or shrubbery in a short time. They sometimes appear in such great numbers that they become dreaded plagues, destroying all the vegetation in their path. Most notorious of these insect herbivores are the locusts and the crickets.

The ROCKY MOUNTAIN LOCUST, a migratory grasshopper, ordinarily remains within a restricted range, but at the onset of hot dry weather they often gather in hordes and sweep over many miles of the grassland biome. During a locust plague of the 1870's, the skies of Nebraska were darkened by a swarm 100 miles wide and 300 miles long, a flying invasion by an estimated 100 billion tiny but devastating herbivores. The MORMON CRICKET, a long-horned black grasshopper, has wings too short for flight and so can cover ground only by crawling and hopping. Countless millions of these crickets periodically leave their homes in the foothills at the western edge of the Great Plains, overrun the grasslands, and consume every bit of vegetation in their path. Metal traps and fences surrounding houses and bordering the roadsides collect them in writhing masses several feet deep and miles in extent. Normally these insects can be kept in check by birds like the western kingbird and scissor-tailed flycatcher, which feed mainly on grasshoppers. Once when the Mormon settlers were subjected to an especially large cricket invasion, a flying army of gulls appeared and waged a successful counteroffensive. This dramatic example of ecological forces maintaining homeostasis in the community is commemorated by a statue in Salt Lake City erected by the grateful Mormons to honor their feathered allies.

The bird herbivores include many small songbirds. They prefer a diet of grass and weed seeds but also eat insects when plant food is not available. A number of these are familiar in fields and open woodlands of the deciduous forest biome: the goldfinches, bobolinks, sparrows, and meadowlarks. With them in the grassland biome is the PRAIRIE CHICKEN, the great game bird of the central states. Its eastern relative, the heath hen, as we have already discovered, succumbed to the advance of civilization along the Atlantic Coast. The prairie chicken is a plump bird, weighing up to two pounds, with brown and black plumage and a short rounded tail. The male sports tufts of erectile black feathers on the neck, and has air sacs on either side of the throat which, when inflated, resemble small oranges. Prairie chickens eat seeds, berries, and small fruits; in summer they devour great quantities of grasshoppers and other insects.

A large number of mammal herbivores have their home in the grasslands, among them a variety of rodents and rabbits. Several kinds of HARVEST MICE feed on the foliage and seeds of the grasses, and weave birdlike nests from their stems and leaves. Harvest mice are small brown rodents with white underparts. They are agile climbers amid their grassy jungles, where they carry on most of their foraging at night. Slightly larger are the DEER MICE, gray or brown in color and similarly lighter on the underside. In addition to grasses and seeds, they eat insects. Deer mice often build nests in burrows abandoned by other animals. The common PRAIRIE VOLE, also small and brown, subsists on leaves, roots, and seeds, often cutting down the taller grasses to reach its meal. This rodent digs its own burrows in the sandy soil of the short-grass prairie. All these mice are active throughout the year.

The easily penetrated soil of the grasslands provides an ideal environment for the many burrowing rodents; by tunneling underground they can find sanctuary when pursued by carnivore enemies. Their activities

PRAIRIE CHICKEN

The prairie dog is an herbivore well adapted for living in the grassland community. (*Credit: Soil Conservation Service, U.S.D.A.*)

benefit the grassland as well as the herbivores themselves; the digging and burrowing mixes the minerals and humus in the soil, and also increases water absorption after heavy rains. The most energetic of these rodent excavators are the pocket gophers, active the year round and capable of digging several hundred feet of tunnel in a day. Of the species living in this biome, typical are the NORTHERN POCKET GOPHER of Montana, Wyoming, and the Dakotas; and the PLAINS POCKET GOPHER, found in the true prairie country of Missouri, Iowa, and Kansas. Pocket gophers are so named because of their unusual fur-lined cheek pockets, which open to the exterior by tiny slits; special muscles enable the gopher to turn these pockets inside out when in need of cleaning. The gopher fills these handy pockets with nuts and seeds, stuffed through the slits with its paws. A pocket gopher is a stout rodent with the habits of a mole, but much larger in size. The small eyes and ears, short neck, and compact body are well suited to a life of burrowing. Its digging tools are powerful incisor teeth and large front claws. This solitary and silent creature rarely ventures far from the opening to its burrow, which may extend for 500 feet and include several cozy bedrooms and storerooms. Pocket gophers vary in color from pale buff to dark brown; often the color is remarkably like that of the soil in which its home is located.

The early explorers of the Great Plains were fascinated by a strange little animal that sat on its haunches, with paws folded on its chest, and barked like a dog. Logically they called this rodent of the grasslands a PRAIRIE DOG. Prairie dogs are plump, sandy-brown animals whose coloring blends with the grasses and the earth mounds surrounding their burrow openings. A full-grown prairie dog may reach a length of thirteen inches and weigh about three pounds. There are two common species. The black-tailed prairie dog is found over a wide range from the Dakotas and Kansas westward across the Great Plains. The white-tailed prairie dog, a more slender animal with a shorter tail, inhabits the mountainous areas of Wyoming, Colorado, Utah, Arizona, and New Mexico. Prairie dogs were formerly much more numerous than at present. Being gregarious animals, they lived in vast underground communities, or "dog towns," which often extended for many miles. One such dog town had an estimated 400 million inhabitants! Prairie dogs are still familiar sights on the western grasslands, but are much reduced in numbers. They are hearty eaters and consume great quantities of grass and similar forage. During the winter months they indulge in long naps. When aboveground they have to keep a sharp lookout for their enemies: birds of prey, coyotes, and badgers.

Other common mammal herbivores of the prairies are the rabbits. The EASTERN COTTONTAIL ranges westward into the grasslands, but the most abundant species is the JACKRABBIT, a larger and heavier animal with a length of twenty-four inches and a weight of eight to ten pounds. The name refers to the huge jackass-like ears, often a third as long as the entire body. They serve as directional air-scoops, enabling the rabbit to detect the slightest sounds. The summer coat is grayish brown, an excellent camouflage among the browning grasses; but in winter, in the northern part of its range, the jackrabbit changes into a white fur coat like the changing coat of the varying hare in the forest biomes. Jackrabbits feed on grasses, but unfortunately like such crops as alfalfa and clover. They seldom drink, obtaining water from their diet of succulent vegetation. In very dry spells when no other plant food is available, they chew off the spines of prickly pear cacti to get at a juicy meal of cactus tissue. By successive leaps which often cover fifteen feet or more of the ground, they can travel as fast as thirty miles per hour, relying on such speed to escape their coyote and fox enemies.

Larger and more spectacular of the herbivores in this biome are the pronghorn antelope and the bison. The PRONGHORN is a graceful, light chestnut-brown and white, hoofed animal, smaller than a deer, usually less than one hundred pounds in weight. The hairs of its white rump patch can be erected to flash a danger signal to other antelopes. Short backward-curving horns, found on both sexes, are black; the covering sheath is shed annually but the bony core is permanent. Pronghorns, swiftest of North American mammals, are able to attain sixty miles an hour in sudden bursts of speed which can easily outdistance enemy coyotes and wolves. The pronghorn is the only ungulate native to our continent; all others have migrated here in prehistoric times via the Bering Strait. Thriving on grasses in the summer, sagebrush in the winter, millions of pronghorns formerly roamed the plains in company with the herds of bison. But agriculture and fenced-in cattle ranges eliminated great areas of their native habitat, and unrestricted hunting also diminished their numbers. By 1908 there were only 19,000 on the entire continent. Finally, in 1953 protective legislation was enacted, and the herds are now estimated to have grown to a third of a million animals. Many are protected in the Hart Mountain National Antelope Refuge in southern Oregon, others in the Sheldon National Antelope Range in northwestern Nevada.

Monarch of the prairies was the mammoth BISON, or buffalo, which has become the symbol of the Old Wild West. A member of the cattle family, a bison is a huge mammal reaching a length of eleven feet, six

Bison herd in a mesquite grove of Wichita Mountains Wildlife Refuge, Oklahoma. (*Credit: Soil Conservation Service, U.S.D.A.*)

feet high at the humped shoulders, and weighing a ton. The fur coat is dark brown in summer, lighter yellowish brown in winter, when the long shaggy fur amply protects the bison from wintry blasts. Both sexes bear small horns, which, as in all cattle, are a permanent horny sheath over a solid bony core. Although their eyesight is poor, the sense of hearing and smell are highly developed. Bison are gregarious animals, roaming the grasslands and woodlands in large herds as they move from one grazing area to another. In winter they brush the snow off the grass by sidewise movements of their oversized heads. Their only enemies, except for man, are wolves and mountain lions, which follow the herds to prey on the sick, the crippled, and the young. Chief pests are the insects that swarm over their bodies; bison gain relief from their attacks by wallowing in sand or mud, leaving the characteristic hollows found in the prairies.

Although hunted by the plains Indians as a source of food, clothing, shelter materials, and other essential items in their lives, the bison population was so great that they were little affected by such hunting. The original bison herds are estimated to have included at least 50,000,000 animals; there may never have been such a large assemblage of mammals elsewhere on the earth. But tremendous numbers were no guarantee of survival when the white man brought changes to the prairie community. Daniel Boone, in 1769, could still find great numbers of bison in Kentucky. The last bison disappeared from the region east of the Mississippi River before the Civil War. Builders of the transcontinental railroads relied on buffalo meat, and indiscriminate "sport" hunting also took place. "Buffalo Bill" Cody himself killed 4280 bison in one year for the railroad crews. The slaughter soon reached a million animals a year. The last bison of the southern herds was shot on the Santa Fé Trail in 1879; the last survivor of the northern herds was killed in North Dakota in 1883. At the turn of the century, as a result, less than 1000 bison were left in the entire United States.

Today the bison is extinct in its wild state, but public opinion and the crusading efforts of such groups as the American Bison Society averted the catastrophe of total extermination. In recent years bison in protected refuges and ranges have been increasing in numbers, with an estimated 6000 now living in the United States. Some thrive on the National Bison Range in northwestern Montana, others in the Wichita Mountains Wildlife Refuge in southwestern Oklahoma. Small herds can be seen in Yellowstone and Wind Cave National Parks. Thus this great American member of the wild cattle tribe, like its close relative the muskox, has been saved by a last moment reprieve from being eliminated from the wildlife community to which it belongs.

The Wildlife Community: The Carnivores

In the carnivore niches of the prairie, the most common animals are the reptiles, the birds, and the mammals. They feed mainly on the numerous rodents and rabbits and so tend to prevent overpopulation of the community by these prolific herbivores. Most of these carnivores also occur in the forest biomes; as a result there are few that are peculiar to the grassland biome.

Among the prairie reptiles are fence lizards, garter snakes, black racers, copperheads, and timber rattlesnakes such as we have already met in the eastern wildlife communities. But a few snakes, both harmless and venomous, are more typical of the grassland biome than of the adjacent wooded areas. All prey on the many kinds of mice, rats, pocket gophers, and other small rodents. The BULLSNAKE is a large but harmless snake found throughout the mixed prairie and the Great Plains. It is a yellowish species with dark brown or black blotches, and grows to a length of six feet. Another harmless snake is the GREAT PLAINS RAT SNAKE, found chiefly in the southern Great Plains region. It is smaller and slimmer, with brown blotches on a gray background. Like the indigo snake of the tropical biome, it makes a good pet because of its gentle disposition.

Two kinds of rattlesnakes live in the grasslands. The MASSASAUGA, found most frequently in the true prairie, is a small spotted snake less than three feet in length; it is sometimes called the "swamp rattler" because of its preference for low wet habitats. The massasauga has a gray ground color with large black spots along the back. More widespread, especially on the Great Plains, is the PRAIRIE RATTLESNAKE, an olive-green snake with dark brown blotches on its back; it grows to a length of four feet. This is the reptile which folklore tells us lives in harmony with prairie dogs in their burrows — a myth not supported by actual fact, since prairie dogs are the snake's favorite diet.

Arch enemies of the small rodents are the birds of prey, predators that can swoop down from the skies and catch unwary animals straying too far from the safety of their burrows. The RED-TAILED HAWK of the eastern woodlands also inhabits the grassland region; from a perch on a lone tree or pole it can scan the ground with keen eyes, and unerringly dive on such prey as lizards, snakes, and mice. The FERRUGINOUS HAWK, a grayish-brown bird with about a five-foot wingspread, flies closer to

BURROWING OWL

the ground, hunting mice, rabbits, and ground squirrels. It is now rare because of persecution by farmers who failed to appreciate its vital role in eliminating excessive numbers of rodents. Two owls of the grasslands also prey on small herbivores and the lesser carnivores. The SHORT-EARED OWL, slightly more than one foot in length, is a buff-colored species whose eartufts are barely visible. It hunts mice ceaselessly, day and night. The short-eared owl rests and sleeps on the ground, protected by its color, which resembles that of the prairie vegetation. Characteristic of the open treeless plains is the small BURROWING OWL, about ten inches long and generally found on the ground. It is a grayish-brown owl with long legs and short tail. For a home it either digs its own burrow or occupies an abandoned gopher hole. Burrowing owls can often be seen sitting at the opening of their underground homes. They are most active in evening and early morning, preying on flying insects, snakes, and small rodents.

Two small mammals occupy the carnivore niche: the ground squirrel and the armadillo. The STRIPED GROUND SQUIRREL, or thirteen-lined ground squirrel, is an omnivore that relies as much upon insects and mice for food as upon grasses, cactus fruits, and other plant materials. Found throughout the biome, this strikingly marked ground squirrel has a distinctive coat of light to dark brown, striped lengthwise with narrow white bands. It is a slender rodent, twelve inches or less in length. The striped ground squirrel, unaffected by noonday heat, is active throughout the day when many prairie animals withdraw to their underground homes. Usually seen sitting upright near the tunnel opening, the small mammal expresses itself in birdlike, high-pitched whistles. In August, well fed and fat, the squirrel retires to its burrow and hibernates for five or six months. Before hibernation, with admirable foresight, it has plunged up all entrances to the retreat, to keep out such unwelcome intruders as weasels and snakes, which would consider the ground squirrel a prize meal.

The most grotesque animal of the biome is the armored tank called an ARMADILLO, member of the primitive and toothless mammal group known as the Edentates, to which the anteater and the sloth also belong. The armadillo is a warm-climate animal that has spread north of the Rio Grande River into the Texas grasslands, and in recent years extended its range east to Florida. Looking like a cross between a small pig and a turtle, this strange mammal has a shell of nine flat bony plates that extend around the body; between the plates grow a few stiff hairs. The pointed snout plows easily through the leafy litter as the armadillo shuffles along the ground in search of insects, which the long sticky tongue sweeps into its mouth. The forefeet possess strong claws with which the animal can dig into the ground. Armadillos dislike the noonday heat and then hide in rocky crevices or underground. Yet the armadillo cannot endure very low temperatures and consequently the northern limits of its range is determined by the occurrence of freezing weather.

Of the larger carnivores, the BADGER, a member of the weasel family, is well adapted for living in both open grasslands and arid deserts. A badger is a squat, bowlegged animal with a grizzled fur coat and black face marked by a white stripe on the forehead and white cheek markings. Full-grown badgers can weigh more than twenty pounds. They are fierce fighters, left severely alone by other carnivores. Their chief claim to distinction is an ability to dig rapidly, aided by the long strong claws on the forelimbs. A badger can dig much faster than a pocket gopher, which it pursues into the ground; when digging to escape an enemy, a badger seems to disappear as if by magic into the earth. In addition to gophers, badgers eat mice and prairie dogs. During the winter months, a badger retreats to its subterranean bedroom for long naps, emerging during warm interludes to search for a meal.

The KIT FOX, also known as the swift fox because of its speed, is an alert, pert, and slender animal. It is the smallest of the fox tribe, being only twenty-four inches long and weighing less than six pounds. A kit fox has grayish to buff-colored fur, very large ears, and a black-tipped

One of the top predators of the grassland biome is the crafty coyote, looking much like a domestic dog. (*Credit: New York Zoological Society*)

tail. It is a nocturnal predator, rarely seen by day. This carnivore is the end of a food chain that includes insects, small reptiles, rodents, and rabbits. Kit foxes range from Canada to Texas, and are found in the southwestern deserts as well.

The top predator of the prairies is the COYOTE, a member of the dog family resembling a sharp-faced collie; the fur coat is grayish or brown, with white underparts and a black-tipped tail. Coyotes attain a maximum weight of fifty pounds. This wild dog of the West can serenade the night with a variety of barks, whines, howls, and wails that sound more like a chorus than a solo performance. The call of the coyote is as much a part of the popular concept of the prairie as the vision of a shaggy bison. Coyotes travel at a steady loping gait, but when necessary can increase their speed to forty miles an hour for short periods; they prefer grassland with some brush or woodland as a protective cover. In recent years coyotes have increased in numbers and extended their range eastward to the Atlantic Coast states. Like the red fox, the coyote is a cunning, adaptable animal which has learned to survive in competition with man. Their usual food consists of mice and ground squirrels, with fruits and berries added in season; they have also become scavengers. When hunting in packs, coyotes attack such large herbivores as antelope and deer.

Man and the Grassland Biome

Both the plant life and the animal life of the grassland biome have been drastically altered by the inevitable need for more agricultural and cattle-raising land. In the true prairie, the physical factors that made this eastern part of the biome a region of tall lush grasses — ample rainfall, level topography, rich soils — also provided an environment for prosperous farms. To make room for them, the native wildlife communities have been eliminated until little remains of the original prairie. This was inevitable, since food requirements of a growing population had to be met. Last-minute attempts, however, are being made to set aside a few wilderness areas where the original communities may survive.

In the western mixed prairie and on the Great Plains, shortsighted farming and grazing policies have in many instances combined with natural causes to bring about large-scale erosion problems. Before the introduction of farming into semi-arid regions unsuited for traditional agricultural methods, and before intensive grazing by domestic cattle occurred on both public and private lands, the ground cover of native grasses protected the soil from erosion. Such vegetation broke the im-

Erosion by dust storms occurs on the prairies when the grass cover is removed. Beaver, Oklahoma. (*Credit: Soil Conservation Service, U.S.D.A.*)

The aftermath of a grassland dust storm at Guymon, Oklahoma. (*Credit: Soil Conservation Service, U.S.D.A.*)

pact of falling rain so that little damage to the soil resulted; the sod and the debris among the grass roots formed a series of dams and terraces holding back the water until it could percolate into the ground. The vegetation also held the surface of the soil in place when high winds might otherwise blow it away.

No other biome has presented a more vivid example of the disastrous results of tampering with an ecosystem than did the events which took place on the western prairies between 1930 and 1940, climaxed by the great dust storm of 1935. After World War I a sudden demand arose for more wheat; to satisfy this need, farmers plowed up thousands of acres of prairie and planted them in wheat. In 1930 the saturation point of wheat production was reached, wheat prices dropped, and farmers abandoned their land. With the original grass cover gone, the loose soil became a potential erosion hazard. During the same year, demand for more meat increased the number of cattle until the herds far exceeded the carrying capacity of the grazing lands. The overgrazed grasslands, denuded of native grasses, also became a latent erosion hazard. Then came a third factor to seal the fate of the land: a great and unprecedented drought, accompanied by high winds. In the abandoned farm areas the soil was picked up and carried aloft. The skies became darkened with dust, as valuable topsoil from one region was dropped on another. Over Wichita, Kansas, the atmosphere to a height of 12,000 feet was laden with 5,000,000 tons of dust carried from lands several hundred miles to the west. The "black blizzard" smothered even the tenacious buffalo grass. During these years, in some states 90 per cent of the protective grass cover was removed. An inch of soil, which required one hundred years to accumulate, was removed by an afternoon of high winds. When rain did come, this amount of soil was washed away by a one-hour shower. Overgrazing, plowing, drought, high winds — together they laid waste many hundreds of square miles of the grassland biome. Prickly pear cactus, Russian thistle, mesquite, and weeds took over the domain of little bluestem, blue grama, and buffalo grass. Finally, in 1941, nature began the cycle of restoration, with several years of above-normal rainfall and lessened winds. Gradually succession began to take place, the buffalo grass returned, and later the other medium- and short-stature grasses.

Droughts and windstorms cannot be prevented, but careless use of the land for either farming or grazing can. If nothing is done to prevent overgrazing in the Great Plains, it may become as disastrous as it proved to be in Texas, where millions of acres denuded of grasses are now mesquite and cactus. Not only did the native vegetation suffer, but also the many animals dependent upon the original grassland environment. The impact of these changes upon the pronghorn and the bison in the prairie biome has already been mentioned. Unnecessary killing of prairie dogs and coyotes, as well as other native animals of the region, is often the result of other poorly conceived policies of land use. All these changes in the grasslands by man have wrought a catastrophic modification of the wildlife community. In no other biome is the importance of ecological knowledge to prevent human mistakes more evident.

13 / The Desert

THE LAST WILDLIFE COMMUNITY we shall explore is in many ways the most fascinating of all. It is the desert biome which extends from the Rio Grande region of Texas to southern California, and northward into the Great Basin between the Rocky Mountains and the Sierra Nevada. Considered a dread no-man's land by most cross-country motorists, an area to be traversed as quickly as possible, the desert is rarely appreciated for what it is: a living laboratory revealing the resourcefulness of life under most trying circumstances. Like the tundra, this biome presents a challenge to life which has resulted in plants and animals with many unique adaptations. Such influence of environment on life has been well expressed by Joseph Wood Krutch.* "We grow strong against the pressure of a difficulty, and ingenious by solving problems. Individuality and character are developed by challenge. And there is no doubt about the fact that desert life has character. Animals and plants, as well as men, become interesting when they fit into their environment.... And nowhere more than in the desert do they reveal it."

The American desert is not a continuous community, with a uniform environment and membership. As in the grasslands, the great north-south extent of the biome creates a wide range of temperature and to some extent, of precipitation. Because of this difference in climate the biome can be divided into a northern and a southern desert, each with its characteristic wildlife. In the *northern desert*, situated in the Upper Sonoran life zone, the precipitation is relatively greater and the temperature lower than in the rest of the biome. Much of the northern community is dominated by sagebrush; xerophytic trees and cacti are less abundant, as is the reptile life. It lies in the Great Basin region of Nevada, Utah, western Wyoming, and southern Idaho and Oregon. In the *southern desert*, located chiefly in the Lower Sonoran life zone, the temperatures are higher and the rainfall more scanty. More xerophytic

* *The Voice of the Desert* (New York: William Sloane Associates, 1955).

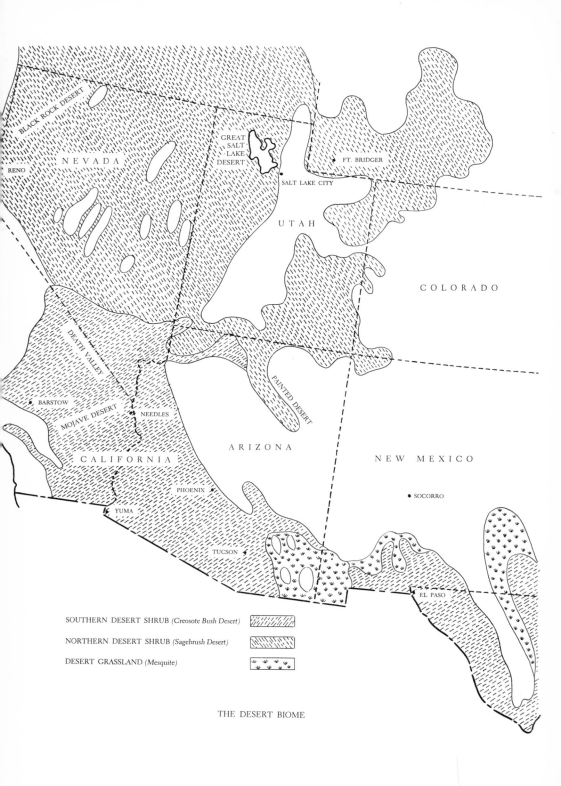

THE DESERT BIOME

plants, such as creosote bush, desert trees, cacti, and other succulents dominate the community; reptile species are varied and numerous. The southern desert includes three distinct regions. The Mojave Desert lies in the southeastern portion of California which borders the northern desert of Nevada; representative of this part of the biome is Death Valley National Monument. The Sonoran Desert is to the south and east of the Mojave Desert, in extreme southern California and Arizona; Saguaro National Monument is located here. The Chihuahuan Desert, lying along the Rio Grande region of Texas, is farthest east, isolated from the rest of the southern desert. Typical of this community is Big Bend National Park.

The American deserts, in terms of geologic time, are of comparatively recent origin. Although they are some five million years old, the deserts came into existence much later than the grasslands of the central United States. Prior to the time when geologic forces changed the climate and thus brought on desert conditions, the southwestern United States was a forested area with a subtropical vegetation. As rainfall decreased and the temperatures became higher, the forests gave way to grassland, and eventually the grassland changed to desert. In the cooler Great Basin area, species of plants and animals from the northern part of North America gradually became adapted to living under the more arid conditions. Farther south, plants and animals from Central and South America migrated into what is now the United States and became established in the southern desert.

The Physical Environment

Groundwater, the most critical factor in a desert environment, is determined chiefly by precipitation. This ranges from ten inches a year in the northern desert to three inches or less in the hottest and driest part of the southern desert. Most comes as rain; snowfall is light except at high elevations in the Great Basin. In the northern desert at Elko, Nevada, the average annual rainfall is nine inches, occurring chiefly in midwinter. It is the same in most of Utah, but more evenly divided throughout the year. In the southern desert, rainfall ranges from eight inches at El Paso to three inches at Yuma. Over most of the Sonoran Desert the rainfall averages five inches, mainly in the winter.

Precipitation, however, is not the only factor that determines the water available for plant and animal life. Low humidity and high evaporation rate reduce the amount actually usable by plants. Since desert rain is likely to come in sudden cloudbursts, much of the water is lost as runoff from the hard, sun-baked surface of the ground, causing flash floods in the periodically dry streambeds. Notwithstanding, enough water does seep into the deeper soil layers so that below six feet some moisture remains at all seasons. This is available to such plants as creosote bush and mesquite whose taproots often descend to depths of thirty feet or more. Plants with surface roots, such as cacti, must develop a spreading network of rootlets to absorb quickly all the water they can, since it remains in the soil for only a short while after every rain.

It may come as a surprise, if you are exploring the desert for the first time, to find that it is not always hot. In winter the difference is often very great between maximum and minimum daily temperatures. It can be a comfortable 75° F. on a January day in Death Valley, yet go below freezing that night. Seasonal variations are also considerable. The northern desert is a region with cold winters, mild to hot summers. In the Great Basin the average temperature on the desert is 28° F. during January, while that of July is 78° F. The southern deserts lie in the temperature belt of cool winters and hot summers. At Palm Springs, California, the January average is 55° F., the July average 90° F. Here the noonday temperature in summer often exceeds 120° F. A temperature so high would be fatal to most of the desert animals if exposed to the direct sun's rays for more than a few minutes.

It may also be a surprise to find that the desert is not one great, flat, sandy plain. The topography is very rugged, with high mountains interspersed with stretches of desert lowland. A characteristic feature of much of the desert landscape is the alluvial fan of erosion sediments which spreads out from the base of the mountains. The northern desert, although in a region called the Great Basin, is not a single large basin but a number of smaller ones separated by mountain ranges covered with a piñon-juniper woodland. The water draining from these mountain slopes fills the basins a few times a year, although, having no outlets, the depressions become alternately temporary salt lakes and alkali flats, like Carson Sink in the Nevada desert. The vegetation in these basins consists of plants that can tolerate such a saline habitat. The soils, especially in the southern desert, become a pavement of pebbles and rock particles above a less compact layer of sand after wind and water have carried away the smaller and lighter particles. This finer material, dispersed by strong winds, creates the frequent sand and dust storms that make travel dangerous in some parts of the southern desert. In many areas the sand is deposited in dunes, as can be seen in the northern part of Death Valley and in White Sands National Monument. Shifting desert sands present a problem to plants in securing a roothold, and make it difficult for burrowing animals to dig their homes.

The Living Environment: The Plant Life

A remarkable feature of the desert vegetation is the variety of life forms, each adapted to a special type of habitat found in the biome. In other biomes, one kind of life form dominates a large area, such as the herbaceous perennial in the grassland, the evergreen woody perennial in the conifer biome, the deciduous perennial in the deciduous forest biome. In the desert, however, we can find growing within the same area — yet not in competition — annuals, herbaceous and woody perennials, deciduous and evergreen shrubs and trees. Each fills its own particular niche, either in habitat or in period of activity. Each meets the problem of water scarcity either by dormancy or by structural modifications.

Desert Annuals. In traveling from Phoenix, Arizona, to Los Angeles we pass through the Sonoran Desert, its arid aspect most evident in the vicinity of the Salton Sea. If we make this trip in summer, autumn, or winter we are impressed by the vast stretches of barren sand and gravel which are visible between the scattered desert shrubs and occasional trees. But if we pass this way in February or March, the lifeless ground is transformed into a continuous and colorful wildflower garden of amazing luxuriance. During the months of December and January the winter rains have brought welcome moisture to the parched soil; the spring months add the necessary warm sunshine. As a result the desert bursts miraculously into life. The millions of seeds which have been lying dormant during the months of heat and drought now sprout rapidly into seedlings. Each tiny plant, compressing its entire life cycle from seed to seed into a few weeks of accelerated activity, develops foliage and flowers in quick succession. There is an urgency in the annual's way of life, brought on by many generations that "know" this is the only way for their kind to survive. The cycle completed, the parent plants die and the desert resumes its barren aspect. But buried in the sand are millions of seeds, waiting to bring rebirth to the desert gardens after next winter's rains. Thus do the short-lived annuals escape the most rigorous period of the desert climate, at the same time bringing periodic beauty and color to the landscape.

The desert wildflower display is a memorable sight. Miles upon miles of roadside sands and stony slopes become carpeted with an array of blooms in a rainbow of colors, contributed by many different species.

Among them are the lavender and pink flowers of VERBENA, the orange-yellow flowers of FIDDLENECK on their curling flower stalks, the purplish blooms of SCORPIONWEED. The large white flowers of the DESERT PRIMROSE open at dusk, to droop the next day soon after sunrise. In some areas, acres of desert resemble a golden-yellow sea where the tiny aster-like flowers of GOLDFIELDS completely cover the ground. Other shades of yellow are contributed by DESERT MARIGOLD, DESERT COREOPSIS, and the showy TIDY-TIPS. These and many more bring the desert to life after the winter rains. Moreover, they add more than color and beauty to the scene; they play an important role in the ecosystem by providing great quantities of seeds and fruits for the desert birds and mammals.

Desert perennials are two types. Some rely upon the deciduous habit, modified leaves, elimination of leaves, or specialized root systems to enable them to increase absorption of water or decrease transpiration. These are the nonsucculent shrubs and trees. Others specialize in water-storage reservoirs in stems or leaves, and these are the succulent perennial, best represented by the cacti and century plants.

Nonsucculent Shrubs. These have become adapted to the desert environment in many ways. Some are evergreen, with hairy or waxy leaves; others bear leaves only during the rainy season; still others have eliminated the leaves entirely. SAGEBRUSH is a densely branched, evergreen shrub usually from two to ten feet high. A member of the chrysanthemum tribe of dicots, it includes a great number of species found throughout western North America. The hairy, wedge-shaped leaves are gray-green, contributing the dominant color to the typical northern desert; minute flowers are borne in silvery-green clusters. The aromatic scent of this characteristic shrub of the "wide open spaces," especially pungent after a sudden shower, is one a traveler through the desert will always remember. Sagebrush is a wildlife food, valuable particularly in winter to herbivores like mule deer, elk, and sage grouse.

PURPLE SAGE, a member of the mint family and thus unrelated to sagebrush, is another aromatic shrub of the northern desert; it is of smaller size, with purplish flowers clustered in slender spikes. Other species grow on dry slopes southward into the Mojave Desert. The rose family adds a member to the desert shrub community in ANTELOPE BRUSH, which grows higher than purple sage but not so tall as sagebrush. Its creamy-white or yellowish flowers grow singly at the ends of short branches. A number of kinds of SALTBUSH, also known as shadscale, grow in the alkali desert basins and the drier sites of the northern desert, as well as in the southern part of the biome. Being members of the pigweed family, these shrubs have inconspicuous flowers; the narrow, oblong leaves are covered with grayish scales, giving a scurfy appearance to the foliage of this desert shrub.

CREOSOTE BUSH is the commonest shrub of the southern deserts, developing extensive communities of evenly spaced plants on rocky desert flats. It is a strongly scented, resinous shrub that grows to a height of nine feet. The small leaves are well protected by a waxy coating against excessive transpiration; amid the foliage appear small yellow flowers, which give rise to cottony seedballs. Creosote bush is an opportunist in its growth habits. On moist slopes it is an evergreen, retaining its foliage throughout the year. On more arid sites the leaves turn brown in summer and the shrub becomes dormant until the rainy season, getting a fresh crop of leaves in January. Desert Indian tribes use the leaves to prepare an antiseptic lotion and the bark to make a glue used in mending pottery. Small mammals, especially ground squirrels, are fond of the seeds.

Two unusual desert shrubs — one a dicot, the other a gymnosperm — have each in its own way solved the problem of excessive transpiration. OCOTILLO is a dicot whose vase-shaped clusters of cane-like stems etch the skyline of rocky ridges of the Sonoran and Chihuahuan Deserts. Each stem, gray-green and leafless for most of the year, is only a few inches in diameter but grows to a height of twelve feet, armed with long spines. After a rainy spell, small leaves appear above the spines, assuring the manufacture of enough food to keep the plant alive until the next rains come. In late spring the tips of the stems bear spires of brilliant red flowers. MORMON TEA, also known as jointfir because of the jointed stems, is a shrubby gymnosperm that has become adapted to desert living. Stiff yellowish stems carry on photosynthesis, substituting for the leaves, which have become reduced to small scales. Thereby this unusual relative of the conifers is guaranteed against possible water loss through the leaves, a method also adopted by the cacti. Mormon tea is common on dry rocky slopes and mesas throughout the entire southern desert.

Desert Trees. A surprising number of angiosperm trees have become adapted to the drought and the heat of the deserts. All are members of the legume, or pea, family and have compound leaves made up of many small leaflets; the fruit is a pod, often with very nutritious beans as seeds. These are eaten by desert herbivores, and are also part of the diet of native Mexicans and Indians. Most of the trees are armed with spines and grow into thorny thickets.

Ocotillo is a tall shrub of the southern deserts with spiny cane-like stems and clusters of small short-lived leaves. Anza Desert State Park, California.

Two kinds of mesquite are common throughout the southern deserts. HONEY MESQUITE, a shrub or low, spreading tree, grows to a height of twenty feet and resembles a spiny dwarfed apple tree. Inch-long spines grow where the leafstalk joins the stem. Mesquite roots penetrate to depths of forty feet or more and thus can obtain water throughout the year. Although this is a deciduous tree, the bright green foliage clothes the branches for most of the year. The ability of the small leaflets, which make up the compound leaves, to fold lengthwise during the daytime reduces transpiration. SCREWBEAN MESQUITE, a slightly larger tree, has grayish-green foliage and twisted pods. Greenish-yellow flowers are borne in compact cylindrical clusters.

The COMMON PALOVERDE is a small tree or shrub, also with yellow flowers, which bears small pods usually constricted between the seeds. The name, of Spanish origin, means "green stick" in reference to the green-barked trunk. The foliage, a grayish green, is absent for most of the year; when the tree is leafless the photosynthetic bark makes enough food to keep the tree alive. This paloverde is frequent on the sides of canyons and on sandy hills; the masses of yellow blossoms are a spectacular spring display in the desert valleys. The YELLOW PALOVERDE of the Chihuahuan Desert is similar in size and shape and, like the common paloverde, lacks the deep root system of the mesquite. Hence these species usually become leafless in summer.

The SMOKE TREE is well named. The ashy-gray trunk and leafless branches, covered in spring with delicate blue blossoms, give an illusive smoky effect to the entire tree, which grows to a height of thirty feet. The short pods bear only a few large seeds. Smoke tree grows in washes and along dry streambeds in the Mojave and Sonoran Deserts. Another shrubby tree of this region (as well as the sandy dunes of Florida) is the CAT'S-CLAW, or Texas mimosa, a very spiny tree often found in thickets together with mesquite. Very small leaflets reduce the transpiration hazard. In spring, cat's-claw, which is related to the ornamental acacias, has somewhat similar yellow blossoms in globular or cylindrical clusters. The flattened seed pods are nutritious, and are ground into a meal by the Indians. Like mesquite, cat's-claw grows throughout the southern desert.

Among the trees without succulent foliage occurs only one of the monocot group, the CALIFORNIA FAN PALM. This tall, slender, fan-leafed palm grows to a height of sixty feet. The upperpart of the trunk is covered by a ragged thatch of old dry leaves, forming a straw-colored skirt that may be of some value in protecting the trunk from intense

The California fan palm survives in a few canyons of the southern desert, such as here in Palm Canyon near Palm Springs, California.

sunlight and heat. The top of the tree is capped by a cluster of huge leaves, reaching a diameter of six feet and supported by spiny leafstalks. This palm is a rare member of the desert community, surviving only in a few canyons where small streams flow, as in Palm Canyon, near Palm Springs. Many million years ago, when the climate was less arid, the California fan palm was much more abundant. Scattered groups remain today along the western and northern borders of the Sonoran Desert. We find these relics of the past in canyons above an old beach line, indicating a prehistoric freshwater lake that once existed in southern California. This desert tree is familiar as an ornamental street tree in the Los Angeles area.

Desert Succulents. The most striking xerophytes of the desert community are those with leaves or stems modified to be water-storage organs. Those with evergreen succulent leaves are represented by the century plants and the yuccas, those leafless but with succulent and enlarged stems are the cacti. The DESERT CENTURY PLANT, or agave,

The desert century plant has a basal cluster of pointed succulent leaves, and produces a tall flowering stalk bearing small yellowish flowers. Anza Desert State Park, California.

develops a ground-hugging rosette of huge sharply pointed leaves that resemble flattened bananas. Year after year, not only is water retained in the thick leaves but food is stored in the short thick stem. When the plant is mature, a flower bud develops, resembling a mammoth asparagus tip; this grows into a flowering stalk frequently nine or ten feet high, terminating in a cluster of yellow flowers. After flowering, the plant dies. Various species grow on the Sonoran and Chihuahuan Deserts. Lechuguilla, a century plant with spiny-edged leaves, often forms a ground-covering thicket in some areas of Big Bend National Park. They require ten to fifteen years (hardly a century!) of preparation to produce the immense flowering stalk.

The Southwest is the home of many curious plants but none more strange than the various species of yucca, members of the lily family with long, sharply pointed leaves. Although succulent, the leaves are not as thick and fleshy as those of century plants. In some species the main trunk is so short that the foliage forms a compact hemispherical mass close to the ground. A central flowering stalk, reaching a height of ten feet, bears a terminal cluster of large creamy-white blossoms. Other species, called SPANISH BAYONET and SPANISH DAGGER, have unbranched trunks five to ten feet tall. The upper portion is covered with an armor of stiff, sword-shaped leaves each twelve to twenty-four inches long. These species are abundant in Big Bend National Park, others are common in the Mojave and Sonoran Deserts. Most grotesque of all the yuccas is the JOSHUA TREE, typical of the Mojave Desert. Growing to a height of thirty feet, the upper part of the trunk develops a tangled mass of branches, clothed with the elongated succulent leaves. The strange name is said to have been given these trees by the Mormons as they made their way through California to Utah; to them the horizontal branches seemed like the outstretched arms of Joshua leading them out of the wilderness to the promised land. A particularly fine forest of these trees has been preserved in Joshua Tree National Monument near Palm Springs, California.

A most successful type of succulent body has been evolved in the cacti; the problem of excessive transpiration has been solved by the elimination of leaves, that of water conservation by the enlargement of the stem to form a living water reservoir. The photosynthetic work ordinarily performed by leaves has been taken over by the green stems. The spines, which protect the succulent stem, have given the cacti an armor as defense against the desert herbivores. All cacti have wide-spreading root systems that absorb great quantities of water during the

A forest of Joshua trees at Joshua Tree National Monument, California.

brief periods of rainfall. Cacti with jointed stems are found in all parts of the biome: prickly pear, cane cactus, and cholla. Also widespread are the cacti with ribbed, unjointed stems: the pincushion, hedgehog, and barrel cacti. Restricted to the Sonoran Desert are the giants of the tribe — the organ pipe cactus and the saguaro.

While driving through the southern desert, wherever we look we can usually see the ubiquitous prickly pears. Some grow in clumps, flattened on the sand, or sprawl over the gravelly soil and rocks of the desert floor. Others are shrubs and small trees, providing shade and shelter for the desert animals. The flowers are large and showy; some yellow, others red or purple. They give rise to spherical or elliptical red fruits or "tunas," a staple diet of native Indians and Mexicans, and many of the desert herbivores as well. The ENGELMANN PRICKLY PEAR, growing frequently among mesquite and paloverde trees, is a shrub with a short main trunk and an intricate tangle of branches covered with long spines. NOPAL PRICKLY PEAR, an even larger shrub, has a trunk fifteen feet high. BEAVERTAIL CACTUS is a common low-growing prickly pear with purplish pads. Although the pads are without spines they are not as harmless as they seem; their velvety surface is covered with tiny bristles that easily break off and become painfully lodged in any animal or person who touches them.

Scattered over the rocky hillsides are thickets of other shrubby species related to the prickly pears, with cylindrical rather than flattened joints, known as cane cacti and chollas. The central stems of CANE CACTUS are slender elongated cylinders supporting a maze of smaller spiny branches. Most profusely armed of all are the chollas. STAGHORN CHOLLA has a central trunk less than an inch in diameter yet attains a height of six feet. This cholla, with antler-like branches, forms extensive thorn forests on rocky slopes in the Sonoran Desert. On exposed gravelly

The ubiquitous prickly pear cacti of the desert are well armed against hungry herbivores.

The cholla cactus forms diminutive forests on rocky slopes throughout the southern deserts.

sites in the hottest part of the desert grows the JUMPING CHOLLA. It has been given this name because the spiny sections are so loosely joined and separate so easily that they seem to jump off the plant. A slight blow anywhere on the cholla will break the sections at the joints and send a shower of them to the ground. Another name for the species is teddy-bear cactus, referring to the golden-yellow spines that thickly cover the stems and outline the plant in a beautiful radiant halo. But on closer inspection we find the beauty quite deceptive. Just brushing against the cactus will transfer some of the living pincushions to our clothing or body.

If we take the time to wander in the desert, we shall discover many small cylindrical or hemispherical cacti well hidden in rock crevices or at the base of boulders. They easily escape notice except when in blossom. Then the conspicuous and colorful flowers attract our attention; frequently they are larger than the plants themselves. These diminutive members of the cactus clan have reduced the plant body to its most compact and effective dimensions, that of a photosynthetic sphere. Smallest are the PINCUSHION CACTI, only a few inches high, their lengthwise ribs edged with lacy spines. Clustered at the base of the parent

Barrel cactus often is rooted in rocky crevices where few other plants can survive. Death Valley National Monument, California.

plant, groups of a dozen or more cacti together form a mound which may have a diameter of several feet. Slightly larger are the erect and cylindrical HEDGEHOG CACTI, growing to a foot in height; they also have the habit of growing together in mound-shaped clusters. Largest of all are the BARREL CACTI, whose lengthwise ribs are protected by stiff curved spines several inches long. Barrel cacti are massive green cylinders, a foot or more in diameter and up to five or six feet tall. This desert sentinel is often seen rooted in a crevice of a high rocky cliff, standing as a lone outpost of life on an otherwise barren skyline.

The vegetation of every biome has its aristocrats: magnolias and tuliptrees in the deciduous forest, redwoods in the conifer forest, royal palms in the tropical biome. Here in the desert the elite are two torch cacti, the organ pipe cactus and the saguaro. Found in Arizona, both are giant columnar plants with fluted sides. ORGAN PIPE CACTUS grows as a cluster of erect trunks arising from a common base. These cylindrical unbranched trunks, each less than a foot in diameter reach skyward to a height of twenty feet. Organ pipe cactus is one of our rarest species, growing on rocky slopes and mesas near the Mexican border. A small forest of this cactus is protected in Organ Pipe National Monument.

Monarch of the southern desert is the SAGUARO. This cactus, fifty feet tall, is a mammoth which exceeds all other succulents in size. Veteran saguaros which have survived drought, windstorms, disease, and human interference live to an age of 200 years. The ribbed trunk, several feet in diameter, produces candelabrum-like branches high above the ground which turn upward to parallel the main trunk. Large waxy-white flowers, at the tips of the branches, open a few at a time during the night. Saguaro fruits split open to reveal a bright red pulp filled with black seeds. Pulp and seeds both are food for desert birds. In order to germinate, a saguaro seed must find a protected spot beneath a desert shrub. There it grows slowly, requiring ten years to become a seedling four inches high. Structurally the saguaro is well designed to

Monarch of the desert is the giant saguaro, protected in the Saguaro National Monument of Arizona.

survive in the desert. The root system forms a water-absorbing network for a radius of sixty feet in all directions; these roots function also as support for the immense, top-heavy column. The pulpy interior of the trunk and branches is supported within a cylindrical basketlike framework of lengthwise wooden ribs. The saguaro has an especially ingenious type of adjustable water-storage system. The accordian-pleated surface expands in diameter to accommodate any increase of water intake, such as occurs after rain. Conversely, the pleats in the surface become deeper during a dry season and so decrease the diameter of the plant, in this way reducing the surface exposed to transpiration. The water-conserving efficiency of the saguaro is strikingly evident when compared with that of a less well adapted desert plant like the date palm. A date palm loses 500 quarts of water per day by transpiration, in comparison with only 1/50 of a quart lost by a twelve-foot saguaro. Even after several rainless years, a saguaro is able to produce flowers and fruit.

This desert giant is vulnerable, however, in several other respects. We have already seen how the saguaro forests have fallen prey to a minute bacterial parasite. Also, the top-heavy trunk is often blown down by high winds if they occur after a rain, when the surface soil is soft and root-anchorage weakened. More serious is the interference by man with the natural balance of the desert community. When he removes the predators by killing snakes and hawks, by exterminating coyotes and foxes with poisoned bait, the rodent population increases tremendously. This in turn affects the saguaro forests, since rodents feed upon the succulent stems of the young cacti. Thus the indirect result of elimination of predators could be extinction of the most prominent member of the biome — the saguaro. Fortunately a large area of the saguaro forest of Arizona lies within the boundaries of Saguaro National Monument, where all precautions are being taken to protect this unique American tree.

The Living Environment: The Herbivores

If we explore the desert only by day, or only in the summer months, we may get the impression that few animals live in this desert community. This apparent lack of animal life is due largely to the nocturnal habits of many species, and to the high percentage of burrowing animals. In the desert 72 per cent of the animal species live underground, in contrast to the forest biomes, where only 6 per cent are burrowing animals. Even those few which are active by day stay mostly in the shade of desert shrubs, venturing into the open only for food or to escape an enemy. The biome is actually the home of a surprising number of herbivores that rely upon seeds and fruits for food as well as upon succulent foliage for their water requirements. Some feed even upon cactus, in spite of the spines, and nibble the foliage of such tough desert plants as yuccas. Plant-eating insects, especially grasshoppers, are seasonally abundant; they occupy an important herbivore role in the pyramid of life on the desert. Upon them many of the smaller reptiles and birds subsist. Among the vertebrates, the most interesting herbivores are the reptiles and mammals.

Reptile Herbivores. In previous biomes the reptiles have been almost exclusively carnivores, but in the desert three common species are vegetarians. These are the crested lizard, the chuckwalla, and the desert gopher tortoise. The colorful CRESTED LIZARD, growing to a length of ten inches, including the long slender tail, can be found on sandy and gravelly ground in both the Mojave and Sonoran Deserts. The name of this handsome lizard refers to a row of enlarged scales along the middle of the back. The grayish-brown body is spotted with darker red markings, and the tail is ringed with dark spots and bands. Crested lizards are exceedingly wary and dash away before we can approach them closely. Their diet consists of leaves, flowers, and fruits, and they

The crested lizard can be found in both the Mojave and Sonoran Deserts. (*Credit: Courtesy of The American Museum of Natural History*)

live in the burrows of small mammals, the openings to whose homes are well concealed beneath a clump of cacti or other shrubs. With the onset of winter they burrow deep into the ground and hibernate there until spring.

If our desert trek takes us to a boulder-strewn hillside in the Mojave or Sonoran Deserts, we may be fortunate enough to catch a glimpse of a fat, dark gray lizard that looks like a fearsome relic of the days of the dinosaurs. This is a CHUCKWALLA, our second-largest lizard, which reaches a length of sixteen inches. The skin, seemingly a few sizes too big for the body, is covered with minute scales; the underside is colored brick-red. Chuckwallas spend most of the day sunning themselves on the rocks; at night, or when pursued, they seek refuge in the nearest crevice. When safely wedged in a fissure of the rock, a chuckwalla inflates its body by forcing extra air into the lungs, thereby making it impossible for an enemy to dislodge it. During the spring and summer chuckwallas grow plump on a diet of leaves, flowers, and fruits. Throughout the winter months they hibernate in underground retreats.

We need keen eyes to detect the camouflaged chuckwalla basking on the rocks, or to glimpse a crested lizard dashing from the shade of one creosote bush to another. But we need not be so alert to discover the

The chuckwalla looks like a relic of the days of the dinosaurs; it lives among boulders on the Mojave and Sonoran Deserts. (*Credit: Courtesy of The American Museum of Natural History*)

The plodding desert tortoise is a reptile herbivore of the southern deserts.

plodding, grayish-brown DESERT TORTOISE slowly making its way over the desert floor in search of succulent grasses, young cacti, or fruits. This weatherbeaten veteran, which lives on all the southern deserts, closely resembles its relative the gopher tortoise of Florida. Old individuals grow to a length of fifteen inches, but today such large specimens are rare; many are unfortunately collected to be taken home as pets, or are run over by automobiles speeding across the desert highways. Desert tortoises can live for many months without drinking, since they obtain the necessary moisture from succulent vegetation. They dig two types of burrows, one for use during the winter, the other as protection against summer heat. The winter burrow is a large community den in which many individuals congregate for their long sleep during the cold months. In spring they awaken and leave the communal burrow, each to go to its own feeding ground. Here the tortoise digs its individual burrow, with a slanting opening that descends to a depth of several feet. In spring and fall, when the days are not excessively hot, the desert tortoise forages during the daylight hours; but the nights are too cool for reptilian comfort so the burrow then becomes a warm bedroom. In summer, on the other hand, the tortoise spends the hot daytime hours in the cool seclusion of its burrow and ventures abroad only at night.

Rodent Herbivores. Among the small mammals of the desert, none are more plentiful and entertaining than the various mice, rats, ground squirrels, and gophers that inhabit the biome. Most of the species occur over a wide range, in both the northern and the southern deserts. Few, however, are seen by the casual traveler, because, unlike the lizards we have met, they have more secretive and nocturnal habits. To get acquainted with them we shall spend a few nights camping on the desert, sleeping under the stars or occupying an abandoned shack. After the sun goes down, we can expect some of the rodents as visitors.

As we light our campfire, tiny white-footed gnomes peer at us out of the darkness, curious to see what is going on. These are DESERT DEER MICE, small rodents with distinctive white underparts and legs; the rest of the body is a sandy brown or gray, which blends into the desert surroundings. They scurry about silently in search of seeds, nuts, or scraps from our meal. They are agile jumpers and nimble climbers, darting into the protection of the thorny undergrowth when alarmed. Deer mice build grass-lined nests in a variety of places: unoccupied gopher burrows, rock crevices, and discarded furniture. They do not hibernate, but are active throughout the year.

Another visitor to our camp is well named. It is the BANNERTAIL KANGAROO RAT, which bounds along on its hind legs like a kangaroo in miniature. Its unusual ability to make the water it needs was described in Chapter 3; this independence of a water supply enables the kangaroo rat to live where most desert animals are at a disadvantage. A kangaroo rat is a friendly, silky-furred animal with an unusually large head and the large black eyes of a nocturnal mammal; it measures twelve to fifteen inches from the tip of the nose to the tuft of hairs which terminates the long slender tail. Cheek slits open into fur-lined pouches, for this "rat" is related to the pocket mice. The unusually long hind legs are used for locomotion, the short and weak forelegs mainly for stuffing seeds, stems, and foliage into the cheek pouches. Kangaroo rats dig their burrows in any kind of soil, piling the excavated material in huge mounds often many feet in diameter and three feet high. The ground beneath is honeycombed with tunnels, storerooms, and bedrooms; each burrow has only one occupant, which never leaves the safety of the underground home until after dark. Many entrances and exits lead to and from the burrows, the openings usually well hidden beneath creosote or mesquite trees. Kangaroo rats, like many rodents, hoard far more food than they can use. In one burrow, fourteen bushels of dried grasses and seeds were stored. Kangaroo rats do not hibernate, but retreat to their cool quarters whenever the summer heat becomes excessive.

If we are using an old shack for shelter, our first night on the desert will be a sleepless one, disturbed by the noisy activities of the champion busybody of all animals, the PACK RAT. Also known as the trade rat and woodrat, this rodent, about thirteen inches in overall length, has gray or buff-colored fur, large ears, and protruding eyes. Inquisitive and impertinent, pack rats are far from furtive in their nocturnal excursions. Most animals confine their activities to securing food, but pack rats forage for much more than seeds, berries, and other plant materials. They have a strange mania for collecting nonedible objects; with noisy abandon they drag about toothbrushes, jewelry, coins, and anything else that strikes their fancy. They will stop carrying one stolen article to pick up another that has just caught their attention. The miscellaneous plunder is taken to their nests, bulky structures of sticks and branches resembling a dry-land beaver lodge. Within it are stored all the stolen treasures, for what reason no one knows. Pack rats seem immune to cactus thorns; they walk over and through spiny chollas, even carrying sections home in their mouths. These they use to line the passageways into their nests, an effective barricade against their carnivore enemies. Bothersome though they may be, pack rats are certainly fascinating desert characters.

During the day we may catch a glimpse of an animal resembling a chipmunk which runs along with bouncing leaps, tail held erect or arched over the back. It is a WHITE-TAILED ANTELOPE SQUIRREL, a vivacious desert midget also known as antelope chipmunk. A stocky rodent with dark back, cinnamon-colored sides, and a narrow white stripe extending from each shoulder to tail, this ground squirrel eats seeds of annuals, succulent tissues of perennials, and especially new growth of young cacti. Like the pack rat, this nimble rodent can walk over cactus spines with impunity. We may be amazed to see it climbing the maze of sharp-tipped yucca leaves to reach the pods, which it rips open to get at the seeds. Its home is in an underground burrow that opens by vertical shafts beneath the protection of a bush. Like many of its relatives, it is the prey of such desert carnivores as rattlesnakes, weasels, coyotes, and foxes.

In the grassland biome we met those interesting rodents the pocket gophers. They have several desert cousins, which live their lives underground, rarely venturing far from their burrows. The WESTERN POCKET GOPHER is a tan or buff-colored animal growing to a length of ten inches. It is highly specialized for digging, and spends most of the waking hours at this occupation. We may see one of these animals stretching its head cautiously out of its burrow, revealing great white incisor teeth that protrude like the tusks of a walrus. These, and the powerful claws on the forefeet, are the pick and shovel of this industrious miner as it tunnels a few inches beneath the surface, and flowers. The dirt they have excavated is bulldozed into a series of mounds outside the openings of lateral passageways from the main tunnels. For three quarters of the year the retiring gopher lies inactive in its snug home, resuming its activity when the rainy season moistens the topsoil and the annuals and perennials upon which it feeds renew their growth.

Desert Rabbits. As we have discovered, we must make the acquaintance of most of the rodent population at night. We may catch sight of rabbits, however, on any daytime ramble in the desert. The large-eared JACKRABBIT of the grasslands is a member of the desert community also; a common species is known as the antelope jackrabbit. These rabbits usually screen themselves effectively behind bushes, but, when flushed, travel in great bounding leaps across the open spaces. Their trails crisscross the sagebrush and creosote bush deserts, wind beneath Joshua trees, circle clumps of cholla. One item in their diet is the tender new growth of yucca leaves, ignored by most desert herbivores. Smaller and less plentiful is the DESERT COTTONTAIL, a desert relative of the common eastern cottontail. It is a grayish animal with a white puffy tail, seeking shelter in any convenient hole in the ground, usually an abandoned ground squirrel's burrow. All rabbits are preyed upon by coyotes, foxes, bobcats, and the larger snakes.

Desert Hoofed Mammals. Two kinds of deer can be seen in the Sonoran Desert region of the Saguaro National Monument. The DESERT MULE DEER resorts to the arid lowlands in winter, to feed on available plant food, especially cactus fruits. In summer it returns to the higher piñon-juniper woodlands. More restricted most of the time to the higher mountainous parts of the desert is the ARIZONA WHITE-TAILED DEER, a smaller member of the deer tribe. Elsewhere in the Southwest the DESERT BIGHORN, an animal closely related to the bighorn sheep we encountered in the Rocky Mountains, was formerly very abundant. It has survived in a few areas, in the desolate and waterless desert mountain ranges, where it lives on grasses, annuals, and foliage of desert shrubs. Desert bighorns have been known to rip open barrel cacti with their horns, to secure the juicy pulp of the thick stems.

Found nowhere else but in the deserts near the Mexican border is our only wild pig, the PECCARY, or javelina. It is a stocky animal three feet in length and weighing as much as fifty pounds; the fur is coarse and grizzled brown. Peccaries have a thick neck, arched back, short legs, and feet provided with cleft hoofs. The upper canine teeth develop into short tusks, useful in plowing up the ground as the grunting, snorting animal searches for roots and bulbs. Peccaries have never learned to economize on water and so stay close to desert streams and seepage areas such as occur in the mesquite thickets along the Rio Grande River. When cornered they fight gamely, even against the jaguar and ocelot, subtropical carnivores that share their habitat.

The Living Environment: The Carnivores

Very few of the large carnivores such as we have met in other biomes are characteristic of the desert, although a few enter the biome from adjacent areas in the grasslands and conifer forests. The smaller, permanent carnivores must be adapted not only to endure the rigorous living conditions of the hot dry climate, but also to detect and capture prey that spend their daylight hours hidden in burrows. Most successful in pursuing the numerous rodents into their retreats are the snakes, and the birds of prey are adept at striking unexpectedly from the air. Smaller birds and mammals eat great quantities of insects, and also at times can feed on unwary herbivores.

Reptile Carnivores. These are the lizards and snakes, both of which reach a great variety and prominence in the desert biome. Carnivorous lizards, like their herbivorous relatives, can occasionally be seen by day, though they usually lie hidden in the shade of desert shrubs or buried in the sand to escape the heat. In the desert food chains, lizards serve as an important check on the insect population, which constitutes a large part of their diet. The most fascinating, as well as the most grotesque, are the HORNED LIZARDS, or "horned toads," of which there are many species. In spite of their forbidding aspect, like that of tiny dinosaurs, they are entirely harmless. The horned lizard has a flattened, squat body covered with spines and a head adorned with a formidable array of larger spines that look like small horns. Our first glimpse of one of these interesting desert inhabitants is of its head barely projecting from the sand in which the rest of the animal is buried. Special valves in the nose keep sand out of the nostrils when the lizard digs through the loose soil. The preferred haunt of this odd reptile is near an anthill. Here it can eat its fill of ants, a favorite meal. As the ants disappear one by one down the horned lizard's throat, the movements of its head and tongue are too rapid for the human eye to follow.

288 / The Biomes of North America

Numerous grayish-brown SPINY LIZARDS, cousins of the eastern fence lizards, scurry along the rocks or scamper through the foliage of yuccas and the spiny branches of cacti. They are adept at catching beetles, flies, and other insects found among the vegetation. More slender, metallic brown WHIPTAIL LIZARDS slink along the ground, their long tails leaving a trail in the sand. When pursued they dig rapidly and soon disappear from sight. The camouflaged OCELLATED SAND LIZARD is remarkably well suited for living on and in sandy habitats: it is able to travel beneath the surface of the sand in a way we can describe only as "sand swimming." Special scales form fringes on the toes, which aid in pushing through the loose sand. Additional adaptations for this unique type of locomotion are fringed eyelids, which close tightly over the eye. Should any sand get beyond this protective shield, the eye fluid encloses the particle in a bubble which is expelled from the corner of the eye. These sand lizards are constantly on the lookout for a variety of insects: flies, ants, beetles, and grasshoppers.

Largest of all the lizards, with a length of twenty-four inches, is the GILA MONSTER, our only poisonous lizard. It is a slow, clumsy reptile with stout body and tail, stubby limbs. The scales are in the form of colored beads, some orange or pinkish, others black; they create a striking mosaic covering the head and back of the lizard. This rare species, today found only in southern Arizona, wanders at dusk through the

The poisonous Gila monster is our largest lizard; it is found today only in southern Arizona. (Credit: *Courtesy of The American Museum of Natural History*)

saguaro forests in search of snakes' or birds' eggs, its principal diet. Gila monsters have venom glands in their lower jaws, but lack the hollow fangs of the rattlesnake. When this lizard bites, the poison is diluted with saliva as it flows into the wound. The menace of the Gila monster has been greatly exaggerated. It is such a sluggish animal that only careless handling can rouse it; therefore, fatalities from its bite have been very few.

Harmless snakes that we may encounter as we explore the desert include the gopher snake, western kingsnake, red racer, and the leaf-nosed snake. Most common is the GOPHER SNAKE, a yellowish snake four feet long, with mottled or checkered markings. This snake eats lizards and smaller snakes, but often prefers a plump rodent which it pursues into its burrow. The WESTERN KINGSNAKE, reaching a length of three feet, is a glossy-black constrictor with narrow white bands across the body. In the grasslands this species is a daytime hunter but in the desert becomes nocturnal in habit. The RED RACER is a more slender, yellowish or reddish-brown snake with black crossbands on the front half of its body. Its diet consists chiefly of insects and rodents. The LEAF-NOSED SNAKE is an unusual species, peculiar in having a flat projection above the nose which may aid it in burrowing into the sand. Being a nocturnal hunter, this snake forages at night for insects and lizard eggs. When approached by an enemy, it hisses and strikes but is entirely harmless. Leaf-nosed snakes are marked with irregular dark blotches, and rarely grow to be more than fifteen inches long.

Poisonous reptiles that do have to be carefully avoided are the numerous rattlesnakes which live in the desert. From the arid grasslands has come the PRAIRIE RATTLESNAKE, ranging throughout the sagebrush as well as creosote bush deserts. From the south-central states the WESTERN DIAMONDBACK RATTLESNAKE roams into the southern desert, from Texas to California. Typical of the Mojave and Sonoran Deserts is the SIDEWINDER, or horned rattlesnake, a small species less than eighteen inches long, creamy gray, with dark blotches along the back. This rattlesnake has developed a special type of locomotion suited to its home in the shifting sands, where to travel in the way of other snakes would be difficult: it is the unusual method by which the sidewinder (another appropriately named reptile!) moves across the sand. The reared head, held at a sharp angle to the body, is projected forward; when the front portion of the snake has landed on the sand, the rest of the body is brought forward through the air in a looping motion and laid down in front of the head. In this manner the snake progresses by a succession of body movements at right angles to its course, never dragging through the sand. The peculiar "horn" over each eye, responsible for one of the common names of this snake, is useful when the snake lies partly buried in the sand; it forms an eyeshade protecting the eye against the bright desert sun, and also as a barrier that keeps sand from drifting into the eyes. Sidewinders feed on lizards and small desert rodents; the other rattlesnakes tackle larger prey, from gophers to rabbits.

Bird Carnivores. The most highly specialized insect-eaters here, as in other biomes, are the woodpeckers. A zebra-backed GILA WOODPECKER, with red patch on the head, is one of several desert woodpeckers that drill into the succulent saguaros, riddling the trunk with holes. The GILDED FLICKER, with yellow wing and tail linings, has the same habit. These woodpecker holes provide daytime shelters for the sparrow-sized ELF OWL, which often can be seen sitting in the opening waiting for dusk, when it sets out to hunt insects and mice. The GREAT HORNED OWL, which we have seen in other biomes, is also at home in the desert, where it is one of the chief feathered predators on the large rodents. This owl often builds a cumbersome nest in the saguaros. Hawks and falcons are, however, the most effective checks on rodent populations. These include the SPARROW HAWK, also found in the central and eastern United States. The RED-TAILED HAWK is another widespread species commonly seen in the desert. It builds a nest of sticks in a saguaro where a branch joins the main trunk. The PRAIRIE FALCON, usually seen on the wing, is the arch enemy of ground squirrels and gophers. When the keen eyes of the falcon detects an unwary rodent far from its burrow, the bird dives like a missile, rarely missing its target. An unusual bird carnivore of the desert is the ROADRUNNER, an entertaining clown that adds humor to the wildlife community. The size of a small chicken, with brown-streaked plumage, a shaggy crest and long pointed beak, and a long tail usually held erect, it is an easily recognized bird. On strong legs it runs over the ground like a small ostrich, the long tail serving as a rudder in tight turns, the flapping wings giving it additional speed. Roadrunners eat anything, from cicadas and grasshoppers to lizards and snakes. A cousin of the eastern cuckoo, this playful bird has earned the name "roadrunner" from its habit of sprinting along highways, especially when pursued by slowgoing vehicles.

Mammal Carnivores. We discover many old friends from other biomes among the few mammal carnivores. From the mountains come bobcats and mountain lions when hungry enough to pursue jackrabbits and ground squirrels. The coyote of the plains has several relatives on

Two unusual and rare carnivores that live at the extreme southern edge of the desert are the ocelot (upper) and jaguar (lower). (*Credit: New York Zoological Society*)

the desert, very similar in appearance. The badger, aided by a keen sense of smell, locates the burrows of kangaroo rats and ground squirrels and in a few moments can dig out the cornered occupant. The inquisitive and friendly little spotted skunk often visits camping parties in the desert. The dainty and gentle little kit fox is another friendly animal often seen around campfires. Two unusual and rare carnivores that live at the extreme edge of the southern desert adjoining Mexico are the ocelot and the jaguar. The OCELOT is a handsome member of the cat tribe with a yellow fur coat, decorated with a variety of black spots, bars, and rings; it reaches a length of four feet, and may weigh as much as forty pounds. It has an unusually gentle disposition, unlike that of most wild members of its family. Ocelots subsist on a varied menu of snakes, rabbits, mice, birds, and an occasional young deer. They were formerly common in the southwestern United States, from southern Arkansas to the Big Bend region of Texas. Today they are sometimes encountered in the thickets along the Rio Grande River. The JAGUAR is the most powerful cat in America, sometimes growing as large as a tiger. Its bright yellow coat is ornamented with large black rosettes. Unusually big individuals reach a length of six feet, weigh 200 pounds or more. Jaguars are adaptable animals, as much at home on the desert as in a tropical jungle. Common in Mexico, the jaguar today is found occasionally in southern portions of Texas, New Mexico, and Arizona. Formerly it roamed as far as California, where the last jaguar was killed in 1860.

Man and the Desert Biome

The inhospitable living conditions of the southern deserts have been a blessing in disguise, as far as the preservation of the wildlife community is concerned. Mining operations have modified a few small areas, and development of desert resorts has had some effect on the surrounding plant and animal life. But outstanding areas — such as the California Joshua tree forests, the Arizona saguaro groves, the Chihuahuan vegetation of the Big Bend region of Texas — are all protected in state and national parks and monuments. Greatest threat has come in the marginal arid grassland portions, where overgrazing has badly disturbed the native vegetation. The desert, however, like the tundra has been spared excessive exploitation as yet. Let us hope that both will continue to be as fortunate in the future.

Each of the great biomes we have explored has its individual personality. Each in its own way reveals how plants and animals, adapted to the same combination of physical conditions and adjusted to a life together in a stable ecosystem, can be molded into one wildlife community. The community, in turn, reveals the impress of the environment in the appearance and habits of its members. It is the end product of various interacting forces sorting out the well adapted from the misfits among the many species attempting to establish themselves in the community.

In the tundra biome life has met the challenge of extreme cold and of the other adverse factors that exist on the arctic frontier. The environment is so rigorous that we wonder how anything as delicate as a living organism can survive. Yet we find there a community made up of many different types of life which have succeeded in meeting this challenge. In the pyramid of life of this biome, lichens and grasses are the successful producers; lemmings and ptarmigans, caribou and muskox, wolves and polar bears the successful consumers — all outstanding examples of adaptation to arctic living conditions.

In the conifer biome the special fitness of the evergreen gymnosperm trees enables them to dominate the producer niche. In the moderately cool environment, the forests of spruce and fir, of pine and redwood provide shelter and food for the arboreal squirrels and birds, the ground-dwelling deer and rabbits, the predatory wolverines and mountain lions that take their place at the consumer level in the pyramid of life of this biome. The best-adapted vertebrates of the conifer forests are the warm-blooded groups, the birds and the mammals.

Farther south in the milder climate of longer summers and shorter winters, the deciduous forest biome is a community where broad-leaved angiosperms are the dominant food producers. In their shelter grow a profusion of smaller plants with edible foliage, fruits, and seeds. Together they support a large consumer population, many of which such as the raccoons and opossums are arboreal, others like the many rodents and rabbits are ground dwellers. The carnivores in the pyramid of life — the bobcats and foxes — have to learn to stalk their prey with patience because of the abundance of protective shelter. The many aquatic habitats, coupled with the warmer climate, permits amphibians and reptiles to join their warm-blooded neighbors in the community.

At the edge of the tropics, the physical environment is conducive to easy survival; warmth and abundant rainfall produces a lush jungle of vegetation to support the consumer levels in the food chain. To find living space some plants, like the mangrove, become adapted to habitats where other vegetation cannot survive; and those like Spanish moss use other plants for support. Here palms become common trees, huge reptiles such as crocodilians become top carnivores. In this biome, survival means capitalizing on adaptations which give an advantage over other living members of the community, rather than meeting the challenge of adverse conditions in the physical environment.

The central lowlands and great plains make up the grassland biome, where lessened precipitation is the critical factor in survival. Here the herbaceous perennial is better fitted to dominate the producer niche than either the evergreen or deciduous tree. A grassland region is a paradise for herbivores. Some, like the prairie dog, seek shelter underground. Others, like the antelope, rely upon speed to escape pursuing carnivores. Symbol of the biome is the bison, once lord of this domain. In the carnivore level of the life pyramid are the fast-flying birds of prey and the crafty coyote. Arboreal mammals and birds are, of course, absent. Amphibians are rare because of their dependence upon water, but reptiles become more common because of their ability to adapt to semi-arid living conditions.

Finally, in the extremely arid Southwest, the desert biome lies in an environment as difficult for life as that of the tundra. As a result it has a similar variety of highly specialized plants and animals, adapted to rigorous living conditions. But here the critical factors are high temperatures and lack of water. The producers in the food chains are mainly xerophytes, of which the saguaro and the Joshua tree are symbols of adjustment to scarcity of water. Cold-blooded animals are even more numerous than warm-blooded ones, and reptiles take over many of the carnivore niches. Rodents are the dominant herbivores, the kangaroo rat representing the acme of animal adaptation to a waterless environment.

We began our venture into ecology by saying "This is the story of the role of environment in determining the appearance, living habits, and distribution of the plants and animals that make up our wildlife heritage." Our exploration of the biomes of North America has presented a wealth of examples to illustrate this truth.

Every kind of organism, from a manatee to a muskox, from a palm to a pine, is the visible expression of adaptation to a particular set of environmental conditions. And as a corollary, it is true that the future existence of such well-adapted organisms depends upon a continuing environment favorable to their existence. Man should remember this wherever and whenever he modifies the conditions that have made the wildlife what it is today and what it can continue to be tomorrow provided that he does not upset the forces maintaining the balance in these wildlife communities.

14 / Wildlife Sanctuaries in North America

IN EACH OF THE BIOMES of North America there are yet areas in which the wildlife community remains to some extent much as it was when the white man came to this continent. These are vignettes, as it were, of the primitive America which is fast disappearing. Such wildlife communities can be found in the public lands that have been set aside as national parks and monuments, national forests, provincial and state parks, wildlife refuges, and game preserves. These publicly owned and controlled lands include millions of acres of tundra, forest, grassland, and desert. Many of these were set aside for other purposes than preservation of wildlife communities; in some, grazing, hunting, trapping, mining, lumbering, and recreation are permitted and thereby have modified the original wildlife community. But in others, the ecosystem has been left undisturbed and thus the original wildlife communities are still present.

Of the thousands of national and state and local parks, and other publicly owned areas, some 500 have been selected which can give us a picture of representative wildlife communities. These include areas large enough to form self-maintaining ecosystems, and to demonstrate the ecological principles underlying our wildlife relations. They provide field laboratories where we can see a biotic community in action. Many of them exist in accessible and well-known national parks, where we can find examples of the biomes and smaller communities described in the previous chapters. Where animal life is not specifically mentioned, it can be considered typical of that part of the biome.

WILDLIFE SANCTUARIES IN ALASKA AND CANADA

I. *The Tundra Biome*

	ACREAGE	VEGETATION	ANIMAL LIFE
1. Aleutian Islands Nat'l Wildlife Refuge (Alaska)	2,700,000	tundra	caribou, wolverine, Alaska brown bear, blue fox, sea otter
2. Katmai Nat'l Monument (Alaska)	2,690,000	tundra, white spruce	brown bear, wolf, moose
3. Kodiak Wildlife Refuge (Alaska)	1,950,000	tundra	Kodiak bear, caribou black-tailed deer, water birds
4. Mt. McKinley Nat'l Park (Alaska)	1,900,000	alpine meadow, black spruce	caribou, Dall sheep, moose, wolf, grizzly bear, black bear, wolverine
5. Nunivak Wildlife Refuge (Alaska)	1,100,000	tundra	reindeer, muskox
6. Kluane Game Sanctuary (Canada)	6,400,000	mountains, glaciers	caribou, Dall sheep, mountain goat, black and grizzly bear, wolf, lynx
7. Peel River Game Preserve (Canada)	4,670,000	tundra, boreal forest	Dall sheep, grizzly bear, wolverine, moose, caribou
8. Mackenzie Mtn. Preserve (Canada)	44,400,000	boreal forest	
9. Yellowknife Preserve (Canada)	44,800,000	tundra, boreal forest	caribou, Arctic fox, grizzly bear, muskox, wolverine, wolf
10. Slave River Preserve (Canada)	1,377,000	delta of Slave River	pine marten, lynx
11. Thelon Game Sanctuary (Canada)	9,600,000	tundra, black spruce, white spruce	caribou, Arctic fox, grizzly bear, muskox, wolverine, wolf
12. Arctic Islands Preserve (Canada)	494,000,000	tundra, ice	caribou, Arctic fox, polar and grizzly bears, muskox, wolverine, wolf
13. Twin Islands Preserve (Canada)	35,200		polar bear

II. *The Conifer Forest Biome*

	ACREAGE	VEGETATION	ANIMAL LIFE
14. Glacier Bay Nat'l Monument (Alaska)	1,165,000	coastal forest	black-tailed deer, mountain goat, wolf, black & grizzly bears, wolverine, coyote, beaver

WILDLIFE SANCTUARIES IN ALASKA AND CANADA

Alaska and Canada (cont'd)

	ACREAGE	VEGETATION	ANIMAL LIFE
15. Glacier Nat'l Park (Canada)	333,000	subalpine & coastal forest	mountain caribou, black & grizzly bears, mountain goat, bighorn
16. Garibaldi Provincial Park (Canada)	288,000	subalpine & coastal forest	grizzly & black bears, mountain goat, bighorn
17. Mt. Revelstoke Nat'l Park (Canada)	64,000	subalpine & coastal forest	mountain caribou, grizzly bear
18. Yoho National Park (Canada)	324,000	Engelmann spruce, lodgepole pine forest	moose, elk, Rocky mountain goat, grizzly and black bears, white-tailed & mule deer, mountain lion
19. Kootenay Nat'l Park (Canada)	375,000	montane forest	moose, elk, mountain goat, grizzly & black bears, cougar
20. Jasper Nat'l Park (Canada)	2,688,000	subalpine & boreal forest	elk, moose, bighorn, mountain goat, mule deer, woodland caribou, black & grizzly bears, mountain lion, wolf, wolverine
21. Banff Nat'l Park (Canada)	1,650,000	subalpine & boreal forest	bighorn, mountain goat, elk, moose, mule deer, black & grizzly bears, wolf, wolverine
22. Waterton Lakes Nat'l Park (Canada)	130,000	subalpine & montane	mule & white-tailed deer, black & grizzly bears, bighorn, mountain goat, elk, mountain lion
23. Wood Buffalo Nat'l Park (Canada)	11,000,000	boreal forest	bison, wolf, black bear mule deer, moose
24. Prince Albert Nat'l Park (Canada)	1,190,000	boreal forest & prairie	moose, elk, white-tailed & mule deer, woodland caribou, black bear, wolf
25. Lac la Ronge Provincial Park (Canada)	729,000	boreal forest	moose, woodland caribou, mule deer, black bear, lynx
26. Riding Mtn. Nat'l Park (Canada)	730,000	boreal forest & prairie	
27. Quetico Provincial Park (Canada)		boreal & deciduous forest	
28. Lake Superior Provincial Park (Canada)	345,000	boreal & deciduous forest	
29. Laurentides Provincial Park (Canada)	256,000	deciduous forest	white-tailed deer, wolf, moose, black bear

Alaska and Canada (cont'd)

	ACREAGE	VEGETATION	ANIMAL LIFE
30. Gaspé Provincial Park (Canada)	224,000	boreal forest	moose, white-tailed deer, woodland caribou, black bear, wolf
31. Fundy Nat'l Park (Canada)	50,000	boreal & deciduous forest	
32. Terra Nova Nat'l Park (Canada)	99,000	boreal & deciduous forest	

WILDLIFE SANCTUARIES IN THE NORTHEASTERN UNITED STATES

	ACREAGE	VEGETATION	ANIMAL LIFE
1. Superior Nat'l Forest (Wilderness Area) (Mich.)	927,000	boreal forest	
2. Isle Royale Nat'l Park (Mich.)	133,000	boreal & deciduous forest	moose, beaver, coyote
3. Chequamegon Nat'l Forest (Mich.)	826,000	deciduous forest	
4. Nicolet Nat'l Forest (Wisc.)	640,000	deciduous forest	
5. Ottawa Nat'l Forest (Mich.)	858,000	deciduous forest	
6. Hiawatha Nat'l Forest (Mich.)	476,000	deciduous forest	
7. Marquette Nat'l Forest (Mich.)	353,000	deciduous forest	
8. Munuscung Bay Waterfowl Refuge (Mich.)	9,800	deciduous & boreal forest	deer, bear, moose, bobcat, waterfowl
9. Wilderness State Park (Mich.)	7,600	second-growth spruce-fir, hemlock-maple	white-tailed deer, black bear, badger
10. Interlochen State Park (Mich.)	278	red & white pine, oak-maple	white-tailed deer
11. Hartwick Pines State Park (Mich.)	8,900	white-red-jack pines, spruce-oak	white-tailed deer, black bear
12. Manistee Nat'l Forest (Mich.)	445,000	deciduous forest	
13. Huron Nat'l Forest (Mich.)	416,000	deciduous forest	
14. Pymatuning State Park (Pa.)	20,000	deciduous forest	white-tailed deer, beaver
15. Allegheny Nat'l Forest (Pa.)	467,000	deciduous forest	
16. Cook Forest State Park (Pa.)	7,300	deciduous forest	white-tailed deer
17. Kitchen Creek Forest (Pa.)	1,200	deciduous forest	
18. McConnell Narrows State Forest (Pa.)	300	deciduous forest	black bear, white-tailed deer
19. Allegheny State Park (N.Y.)	58,000	deciduous forest	black bear, white-tailed deer
20. Adirondack Forest Preserve (N.Y.)	2,250,000	deciduous & boreal forest	white-tailed deer, black bear, bobcat
21. Catskill Forest Preserve (N.Y.)	235,000	deciduous forest	white-tailed deer, black bear, bobcat
22. Bear Mtn. State Park (N.Y.)	4,800	deciduous forest	white-tailed deer
23. Green Mtn. Nat'l Forest (Vt.)	230,000	deciduous forest	white-tailed deer

WILDLIFE SANCTUARIES IN THE NORTHEASTERN UNITED STATES

Northeastern United States (cont'd)

	ACREAGE	VEGETATION	ANIMAL LIFE
24. Mt. Greylock State Reservation (Mass.)	9,000	deciduous forest, some boreal forest	
25. Kent Falls State Park (Conn.)	275	white pine-hemlock-maple climax	
26. Macedonia Brook State Park (Conn.)	1,800	northern hardwood forest	
27. Pisgah Mtn. Tract (N.H.)	20	virgin white pine	
28. Rhododendron State Reservation (N.H.)	300	hemlock-rhododendron	
29. White Mtn. Nat'l Forest (N.H.)	674,000	alpine tundra, boreal & deciduous forest	white-tailed deer, black bear, bobcat
30. Baxter State Park (Maine)	193,000	alpine tundra, boreal & deciduous forest	white-tailed deer, caribou, black bear, bobcat, lynx, moose
31. Acadia Nat'l Park (Maine)	30,000	boreal & deciduous forest	white-tailed deer

WILDLIFE SANCTUARIES OF THE MIDEASTERN UNITED STATES

	ACREAGE	VEGETATION	ANIMAL LIFE
1. Mississippi Palisades State Park (Ill.)	1,100	open woodland, prairie	
2. White Pines Forest State Park (Ill.)	380	white pine forest	
3. Black Hawk State Park (Ill.)	200	deciduous forest	
4. Starved Rock State Park (Ill.)	1,400	oak-hickory forest	
5. Cook County Forest Preserve (Ill.)	35,000	second-growth oak-hickory-maple	
6. Piatt County Forest Preserve (Ill.)	550	second-growth deciduous forest	
7. Brownfield Woods (Ill.)	60	nearly virgin oak-maple	
8. Allerton Park (Ill.)	1,400	nearly virgin oak-hickory	
9. Spitler Woods State Park (Ill.)	200	deciduous woods	
10. Pere Marquette State Park (Ill.)	5,100	deciduous woods	
11. Fox Ridge State Park (Ill.)	700	deciduous woods	
12. Giant City State Park (Ill.)	1,500	deciduous woods	
13. Crab Orchard Nat'l Wildlife Refuge (Ill.)		bottomland forest	white-tailed deer, beaver
14. Cave-in-Rock State Park (Ill.)	65	deciduous woods	
15. Shawnee Nat'l Forest (Ill.)	210,000	bottomland forest	
16. Indiana Dunes State Park (Ind.)	2,100	oak-hickory; dune succession	
17. Pokagon State Park (Ind.)	1,000	oak-hickory; tamarack swamps	white-tailed deer
18. Shades State Park (Ind.)	2,000	oak-hickory; tamarack swamps	white-tailed deer
19. Turkey Run State Park (Ind.)	1,500	nearly virgin beech-maple, oak-hickory	
20. Brown County State Park (Ind.)	17,000	second-growth beech-maple, oak-hickory	white-tailed deer
21. Versailles State Park (Ind.)	5,700	partly virgin oak-hickory, sweetgum, red maple	
22. Clifty Falls State Park (Ind.)	700	partly virgin oak-hickory beech	
23. Spring Mill State Park (Ind.)	1,100	partly virgin tuliptree, beech, oak	

Mideastern United States (cont'd)

	ACREAGE	VEGETATION	ANIMAL LIFE
24. Indiana Pioneer Mothers Memorial Forest (Ind.)	250	partly virgin beech-maple	
25. Hoosier Nat'l Forest (Ind.)	117,000	Ohio River bottomlands	
26. Celeryville Bog (Ohio)	600	disturbed deciduous forest	
27. Cleveland Metropolitan Parks (O.)	1,200	partly virgin oak-hickory, beech-maple	
28. Western Reserve University Area (O.)	100	disturbed, beech-maple	
29. City of Akron Reservation (O.)	1,500	floodplain, gorge; beech-maple, hemlock	
30. Mill Creek City Park (O.)	1,400	deciduous & boreal forest transition	
31. Glen Helen Park (O.)	900	second-growth bottomland forest	
32. Cedar Bog (O.)	400	partly disturbed forest	
33. Ohio State University Area (O.)	4,300	second-growth forest	
34. Rock House State Forest Park (O.)	500	mesophytic hemlock forest	
35. Ash Cave State Forest Park (O.)	200	virgin deciduous forest-boreal forest transition	
36. Cantville Cliff State Forest Park (O.)	300	partly disturbed forest	
37. Conklis Hollow State Forest Park (O.)	600	second-growth forest	
38. Old Man's Cave State Forest Park (O.)	1,300		
39. Wayne Nat'l Forest (O.)	106,000	Ohio River bottomland forest	
40. Heuston Woods (O.)	200	second-growth beech-maple	
41. Mount Airy City Forest (O.)	1,100	mesophytic forest, formerly grazed	
42. University of Cincinnati Area (O.)	68	partly disturbed forest	
43. Kentucky Woodlands Nat'l Wildlife Refuge (Ky.)		on Tennessee River	waterfowl
44. Pennyrile Forest State Park (Ky.)	14,000	southern Appalachian forest	
45. Audubon State Park (Ky.)	500	virgin beech forest	
46. Mammoth Cave Nat'l Park (Ky.)	50,000	second-growth southern Appalachian forest	white-tailed deer, beaver, wild turkey

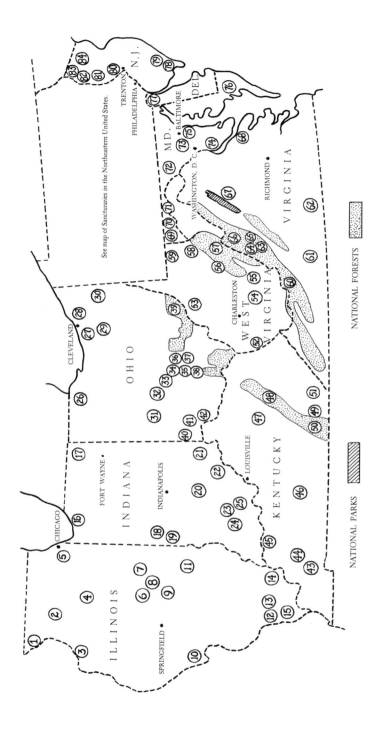

Mideastern United States (cont'd)

	ACREAGE	VEGETATION	ANIMAL LIFE
47. University of Kentucky Area	15,000	second-growth forest	
48. Natural Bridge State Park (Ky.)	1,300	second-growth oak-chestnut oak-pine	
49. Cumberland Falls State Park (Ky.)	1,000	disturbed oak-pine, hemlock-beech	
50. Cumberland Nat'l Forest (Ky.)	450,000	oak-chestnut, oak-pine	
51. Pine Mtn. State Park (Ky.)	3,100	partly virgin oak-chestnut	
52. Cabwaylingo State Forest (W.V.)	7,400	southern Appalachian deciduous forest	
53. Kanawha State Forest (W.V.)	6,700	,,	
54. Babcock State Park (W.V.)	3,200	,,	
55. Watoga State Park (W.V.)	10,000	,,	
56. Holly River State Park (W.V.)	7,500	,,	
57. Monongahela Nat'l Forest (W.V.)	805,000	,,	
58. Blackwater Falls State Park (W.V.)	1,600	,,	
59. Coopers Rock State Forest (W.V.)	13,300	,,	
60. Jefferson Nat'l Forest (Va.)	590,000	southern Appalachian deciduous forest	
61. Fairystone State Park (Va.)	4,500	partly virgin deciduous forest	
62. Staunton River State Park (Va.)	700	disturbed deciduous forest	white-tailed deer, turkey, otter
63. Ramsey's Draft Natural Area (Va.)	1,700	nearly virgin hemlock-oak-chestnut	
64. Douthat State Park (Va.)	4,400	oak-chestnut-pine forest	white-tailed deer, bobcat, turkey
65. Little Laurel Run Natural Area (Va.)	2,000	nearly virgin oak-chestnut, hemlock-pine	black bear, white-tailed deer
66. George Washington Nat'l Forest (Va.)	820,000	Blue Ridge Mtns. forest	black bear, white-tailed deer
67. Shenandoah Nat'l Park (Va.)	182,000	second-growth oak-chestnut; crest of Blue Ridge Mtns.	white-tailed deer, bobcat
68. Westmoreland State Park (Va.)	1,300	second-growth beech	
69. Savage River State Forest (Md.)	52,000	southern Appalachian deciduous forest	

(For Pennsylvania see chart of the Northeastern United States.)

Mideastern United States (cont'd)

	ACREAGE	VEGETATION	ANIMAL LIFE
70. Potomac State Forest (Md.)	12,000	southern Appalachian deciduous forest	
71. Green Ridge State Forest (Md.)	25,000	"	
72. Cunningham Falls State Park (Md.)	4,400	northern Blue Ridge forest	
73. Patuxent Research Refuge (Md.)	2,600	second-growth oak-beech, tuliptree, sweetgum, southern pines	
74. Plummer's Island Preserve (Md.)	58	partly virgin deciduous forest	
75. Patapsco State Park (Md.)	5,000	deciduous forest	
76. Pocomoke State Forest (Md.)	12,000	coastal woodland	
77. Brandywine City Forest (Del.)	2,000	oak-chestnut, partly disturbed	
78. Belleplain State Forest (N.J.)	6,400	deciduous forest	
79. Brigantine Nat'l Wildlife Refuge (N.J.)			Atlantic Coast water birds
80. Washington Crossing State Park (N.J.)	370		
81. Hacklebarney State Park (N.J.)	200	mountain deciduous forest	
82. Swartswood State Park (N.J.)	700	"	
83. Stephens State Park (N.J.)	200	"	
84. Ringwood Manor State Park (N.J.)	600	"	

WILDLIFE SANCTUARIES IN THE SOUTHEASTERN UNITED STATES

	ACREAGE	VEGETATION	ANIMAL LIFE
1. Reelfort Lake State Park (Tenn.)	100	southern Appalachian deciduous forest	
2. Reelfoot Wildlife Refuge (Tenn.)	9,200	"	waterfowl, muskrat, mink
3. Reelfoot Lake State Game Preserve (Tenn.)	44,000	cypress swamps	
4. Natchez Trace State Park (Tenn.)	1,200	disturbed deciduous forest	
5. Chickasaw State Park (Tenn.)	11,000	central lowlands deciduous forest	
6. Montgomery Bell State Park (Tenn.)	3,700	"	
7. Cedars of Lebanon State Park (Tenn.)	8,300	red cedar woodland	
8. Standing Stone State Park (Tenn.)	8,700	central lowlands deciduous forest	
9. Pickett State Park (Tenn.)	1,200	Cumberland Plateau deciduous forest	
10. Cumberland State Park (Tenn.)	1,500	"	
11. Fall Creek Falls State Park (Tenn.)	15,000	"	
12. Harrison Bay State Park (Tenn.)	1,500	Great Smoky Mtns. forest	
13. Cove State Park (Tenn).	850	Cumberland Plateau deciduous forest	
14. Great Smoky Mtns. Nat'l Park (Tenn.; also N.C.)	456,000	half virgin oak-chestnut, boreal forest of red spruce	black bear, bobcat
15. Big Ridge State Park (Tenn.)	3,600	Cumberland Plateau, Great Smoky Mtns. forest	
16. Cherokee Nat'l Forest (Tenn.)	594,000	Great Smoky Mtns. forest	
17. Pisgah Nat'l Forest (N.C.)	479,000	southern Appalachian mountain forest	
18. Mt. Mitchell State Park (N.C.)	1,200	highest summit east of Mississippi River	
19. Nantahala Nat'l Forest (N.C.)	447,000	southern Appalachian mountain forest	
20. Hanging Rock State Park (N.C.)	3,800	Blue Ridge Mtns. deciduous forest	
21. Duke University Area (N.C.)	5,000	second-growth deciduous forest	
22. Uwharrie Nat'l Forest (N.C.)	43,000	oak-pine piedmont forest	

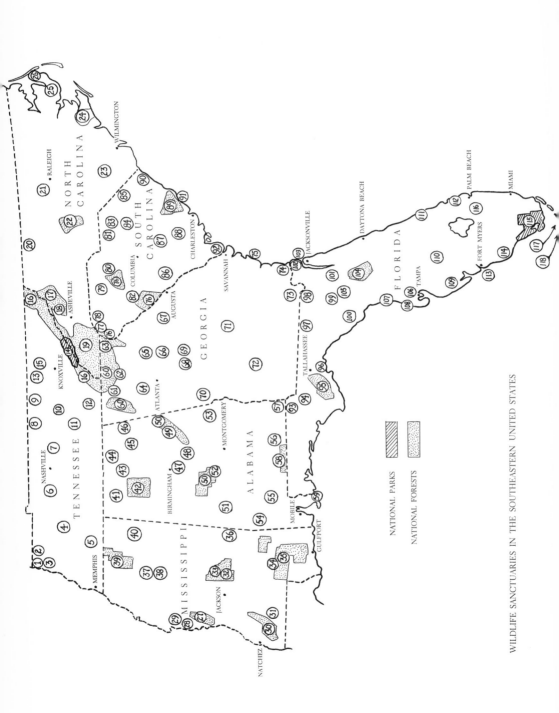

WILDLIFE SANCTUARIES IN THE SOUTHEASTERN UNITED STATES

Southeastern United States (*cont'd*)

	ACREAGE	VEGETATION	ANIMAL LIFE
23. Jones Lake State Park (N.C.)	2,000	river-bottomland forest	
24. Croatan Nat'l Forest (N.C.)	152,000	coastal cypress swamps	
25. Mattamuskeet Wildlife Refuge (N.C.)	50,000	loblolly pine, coastal swamps	waterfowl
26. Pea Island Wildlife Refuge (N.C.)	5,800	coastal sand dune vegetation	waterfowl
27. Delta Nat'l Forest (Miss.)	140	oak-hickory-gum	
28. Yazoo Migratory Waterfowl Refuge (Miss.)		Mississippi River	waterfowl
29. Leroy Percy State Park (Miss.)	2,400	Mississippi River bottomland forest	
30. Homochitto Nat'l Forest (Miss.)	189,000	oak-hickory, Mississippi River bottomland forest	
31. Percy Quin State Park (Miss.)	1,600	southern oak-hickory forest	
32. Roosevelt State Park (Miss.)	560	in Bienville Nat'l Forest	
33. Bienville Nat'l Forest (Miss.)	175,000	coastal-plain forest	
34. Shelby State Park (Miss.)	800	coastal-plain forest	
35. De Soto Nat'l Forest (Miss.)	500,000	coastal-plain forest	
36. Clarkco State Park (Miss.)	800	coastal-plain forest	
37. Carver Point State Park (Miss.)	400	oak-hickory forest	
38. Hugh White State Park (Miss.)	700	oak-hickory forest	
39. Holly Springs Nat'l Forest (Miss.)	123,000	oak-hickory, Mississippi River lowlands forest	
40. Tombigbee State Park (Miss.)	800	central lowlands deciduous forest	
41. Joe Wheeler State Park (Ala.)	2,200	edge of Cumberland Plateau forest	
42. William Bankhead Nat'l Forest (Ala.)	178,000	edge of Cumberland Plateau forest	
43. Wheeler Nat'l Wildlife Refuge (Ala.)		"	waterfowl
44. Monte Sano State Park (Ala.)	2,000	red cedar woodland	
45. Little Mtn. State Park (Ala.)	4,000	Cumberland Plateau forest	

Southeastern United States (cont'd)

	ACREAGE	VEGETATION	ANIMAL LIFE
46. De Soto State Park (Ala.)	4,800	southern Appalachian deciduous forest	
47. Oak Mtn. State Park (Ala.)	10,000	edge of southern Appalachian forest	
48. Weogufka State Forest (Ala.)	800	"	
49. Cheaha State Park (Ala.)	2,700	second-growth oak-hickory	
50. Talladega Nat'l Forest (Ala.)	357,000	oak-hickory forest	
51. Chickasaw State Park (Ala.)	560	coastal-plain forest	
52. Valley Creek State Park (Ala.)	1,000	adjacent to Talladega Nat'l Forest	
53. Chewacla State Park (Ala.)	800	coastal-plain forest	
54. Bladon Springs State Park (Ala.)	350	bottomland forest	
55. Little River State Forest (Ala.)	2,100	bottomland forest	
56. Geneva State Forest (Ala.)	7,100	coastal-plain forest	
57. Chattahoochee State Park (Ala.)	600	bottomland forest	
58. Conecuh Nat'l Forest (Ala.)	83,000	coastal-plain forest	
59. Gulf State Park (Ala.)	5,600	shore of Gulf of Mexico	
60. Chattahoochee Nat'l Forest (Ga.)	665,000	southern Blue Ridge Mtns. forest	
61. Fort Mtn. State Park (Ga.)	2,500	southern Appalachian deciduous forest	
62. Amicalola Falls State Park (Ga.)	240	"	
63. Black Rock Mtn. State Park (Ga.)	1,100	in Nantahala Nat'l Forest	
64. George W. Carver State Park (Ga.)	350	southern Appalachian deciduous forest	
65. Fort Yargo State Park (Ga.)	1,500	piedmont oak-pine forest	
66. Hard Labor Creek State Park (Ga.)	5,800	"	
67. Stephens State Park (Ga.)	1,100	"	
68. Indian Springs State Park (Ga.)	600	"	
69. Piedmont Nat'l Wildlife Refuge (Ga.)		"	

Southeastern United States (cont'd)

	ACREAGE	VEGETATION	ANIMAL LIFE
70. F.D.R. State Park (Ga.)	5,000	piedmont oak-pine forest	
71. Little Ocmulgee State Park (Ga.)	1,400	coastal-plain forest	
72. Chehaw State Park (Ga.)	600	coastal-plain forest	
73. Okefenokee Nat'l Wildlife Refuge (Ga.)	328,000	cypress swamps, upland coastal-plain forest	white-tailed deer, black bear, mountain lion, alligator, water birds
74. Crooked River State Park (Ga.)	600	coastal-swamp forest	
75. Blackbeard Island Wildlife Refuge (Ga.)	4,800	cabbage palm–magnolia coastal swamp	white-tailed deer, alligator, otter
76. Sumter Nat'l Forest (S.C.)	341,000	piedmont oak-pine forest	
77. Oconee State Park (S.C.)	1,100	southern Appalachian deciduous forest	
78. Pleasant Ridge State Park (S.C.)	300	"	
79. Croft State Park (S.C.)	7,100	piedmont oak-pine forest	
80. Chester State Park (S.C.)	500	"	
81. Carolina Sandhills Wildlife Refuge (S.C.)		"	
82. Greenwood State Park (S.C.)	900	edge of Sumter Nat'l Forest	
83. Cheraw State Park (S.C.)	7,300	piedmont oak-pine forest	
84. Lee State Park (S.C.)	2,800	coastal-plain forest	
85. Little Pee Dee State Park (S.C.)	800	coastal-plain forest	
86. Aiken State Park (S.C.)	1,000	coastal-plain forest	
87. Santee State Park (S.C.)	2,300	coastal plain–swamp forest	
88. Givhans Ferry State Park (S.C.)	1,200	coastal plain–swamp forest	
89. Francis Marion Nat'l Forest (S.C.)	245,000	coastal plain–swamp forest	
90. Myrtle Beach State Park (S.C.)	300	coastal forest, beach vegetation (yucca, cactus)	
91. Cape Romain Nat'l Wildlife Refuge (S.C.)		coastal portion, Francis Marion Nat'l Forest	waterfowl

Southeastern United States (cont'd)

	ACREAGE	VEGETATION	ANIMAL LIFE
92. Savannah Nat'l Wildlife Refuge (S.C.)		coastal swamp forest	waterfowl
93. Florida Caverns State Park (Fla.)	1,100	nearly virgin hammock, southern pines	fox, bobcat
94. Torreya State Park (Fla.)	1,000	virgin magnolia-beech forest; rare Torreya gymnosperm tree	white-tailed deer, turkey, bobcat, mountain lion
95. Apalachicola Nat'l Forest (Fla.)	556,000	coastal-plain forest	
96. St. Marks Nat'l Wildlife Refuge (Fla.)		edge of Apalachicola Nat'l Forest	waterfowl
97. Suwannee River State Park (Fla.)	1,800	nearly virgin coastal deciduous forest	
98. Osceola Nat'l Forest (Fla.)	157,000	edge of Okefenokee Swamp area	
99. O'Leno State Park (Fla.)	1,300	coastal-plain pine forest	white-tailed deer, bobcat, fox
100. Manatee Springs State Park (Fla.)	1,600	on Suwannee River; swamp forest	manatee (now rare)
101. Gold Head Branch State Park (Fla.)	1,300	southern pine–scrub oak	alligator, bobcat
102. Fort Clinch State Park (Fla.)	1,000	nearly virgin coastal oak forest	
103. Little Talbot Island State Park (Fla.)	2,500	coastal vegetation, mouth of St. Johns River	
104. Ocala Nat'l Forest (Fla.)	360,000	southern pines–cabbage palm–cypress	turkey, bobcat, white-tailed deer
105. Austin Carey Forest (Fla.)	2,000	nearly virgin pine forest	white-tailed deer, bobcat
106. Hillsborough River State Park (Fla.)	2,800	nearly virgin pine–cabbage palm forest	white-tailed deer, black bear, bobcat
107. Chassahowitzka Nat'l Wildlife Refuge (Fla.)	2,000	Gulf of Mexico, coastal forest oak–pine	white-tailed deer, turkey, bobcat
108. Anclote Nat'l Wildlife Refuge (Fla.)		"	
109. Myakka River State Park (Fla.)	26,000	saw-palmetto prairie and hammock swamp	alligators, cattle egrets
110. Highlands Hammock State Park (Fla.)	3,800	hammock; cabbage palm–pine-cypress forest	white-tailed deer, turkey, alligator
111. Pelican Island Nat'l Wildlife Refuge (Fla.)		coastal-beach vegetation	waterfowl

Southeastern United States (cont'd)

	ACREAGE	VEGETATION	ANIMAL LIFE
112. Jonathan Dickinson State Park (Fla.)	9,500	coastal-beach vegetation	
113. Sanibel Nat'l Wildlife Refuge (Fla.)		subtropical palm forest	waterfowl
114. Collier-Seminole State Park (Fla.)	6,400	mangrove & cypress forest; royal palm hammock	white-tailed deer, black bear, mountain lion, turkey, alligator
115. Everglades Nat'l Park (Fla.)	1,258,000	mangrove swamps, subtropical & tropical hammocks, sawgrass marshes	manatee, crocodile, alligator, white-tailed deer, mountain lion, egrets, ibis
116. Loxahatchee Nat'l Wildlife Refuge (Fla.)		Okeechobee marshland	waterfowl
117. Great White Heron Nat'l Wildlife Refuge (Fla.)		Florida Keys	great white heron
118. Nat'l Key Deer Refuge (Fla.)	1,000	Florida Keys	key deer

WILDLIFE SANCTUARIES IN THE CENTRAL UNITED STATES

	ACREAGE	VEGETATION	ANIMAL LIFE
1. Black Coulee Nat'l Wildlife Refuge (Mont.)	1,500	shortgrass prairie	sage hen, antelope
2. Hewitt Lake Nat'l Wildlife Refuge (Mont.)	1,400	shortgrass prairie	sage hen, antelope
3. Fort Peck Game Range (Mont.)	946,000	shortgrass prairie	mule deer, white-tailed deer, antelope, bobcat, prairie dog, coyote
4. Central Plains Experimental Range (Colo.)	10,000	shortgrass prairie	antelope, mule deer, coyote, badger, prairie dog
5. Upper Souris Nat'l Wildlife Refuge (N.D.)	32,000	tallgrass prairie	grouse, prairie chicken
6. Lower Souris Nat'l Wildlife Refuge (N.D.)	58,000	mixed prairie; bur oak sandhills	white-tailed deer, coyote, red fox, badger
7. Theodore Roosevelt Wildlife Refuge (N.D.)	12,000	mixed prairie & badlands	antelope, coyote, white-tailed deer
8. Wells County Game Refuge (N.D.)	640	shortgrass prairie	white-tailed deer, coyote, red fox, badger
9. Theodore Roosevelt Nat'l Memorial Park (N.D.)	68,000	badlands	
10. Morton County Refuge (N.D.)	600	shortgrass prairie	
11. Dawson Refuge (N.D.)	2,500	shortgrass prairie	white-tailed deer, grouse
12. Long Lake Nat'l Wildlife Refuge (N.D.)	19,000	shortgrass prairie	waterfowl, grouse
13. Badlands Nat'l Monument (S.D.)	150,000	shortgrass prairie	badger, coyote, bobcat, prairie dog
14. Waubay Nat'l Wildlife Refuge (S.D.)		tallgrass prairie	waterfowl
15. Lake Andes Nat'l Wildlife Refuge (S.D.)		tallgrass prairie on Missouri River	waterfowl
16. Middle River State Park (Minn.)	285	prairie-forest	
17. Pipestone Nat'l Monument (Minn.)	276	virgin tallgrass prairie	
18. Chippewa Nat'l Forest (Minn.)	637,000	eastern edge of deciduous forest	
19. Superior Nat'l Forest (Minn.)	1,956,000	deciduous & boreal forest transition along Lake Superior	

Central United States (cont'd)

	ACREAGE	VEGETATION	ANIMAL LIFE
20. Wild Cat Range Game Preserve (Nebr.)	1,000	shortgrass prairie	bison
21. Crescent Lake Nat'l Wildlife Refuge (Nebr.)	46,000	mixed prairie, sandhills	antelope
22. Fort Niobrara Nat'l Wildlife Refuge (Nebr.)	18,000	mixed prairie, dunes, lakes	waterfowl, elk, bison, mule deer, white-tailed deer
23. Valentine Wildlife Refuge (Nebr.)	69,000	mixed prairie	waterfowl
24. Nebraska Nat'l Forest (Nebr.)	206,000	mixed subclimax deciduous forest	
25. Fontenelle Forest (Nebr.)	300	oak-hickory & prairie transition	
26. Gitchie Manitou Monument (Iowa)	200	tallgrass prairie	
27. Silver Lake Fen (Iowa)	28	tallgrass prairie	
28. Lakeside Praire (Iowa)	60	tallgrass prairie	
29. Pilot Knob State Park (Iowa)	370	tallgrass prairie	
30. Lime Springs Prairie (Iowa)	240	tallgrass prairie	
31. Oak Grove Recreational Preserve (Iowa)	100	prairie & woodland	white-tailed deer, red & gray foxes
32. Stone State Park (Iowa)	900	prairie & woodland	
33. Pilot Mound (Iowa)	40	tallgrass prairie	
34. Waubonsie State Park (Iowa)	730	prairie & woodland	
35. Kirwin Nat'l Wildlife Refuge (Kan.)		tallgrass prairie	waterfowl
36. Washington Marlatt Memorial Park (Kan.)	160	tallgrass prairie	
37. Fort Hays Branch Experiment Station (Kan.)	7,600	mixed prairie	
38. Sawn Lake Nat'l Wildlife Refuge (Mo.)		deciduous woodland	
39. Mark Twain Nat'l Forest (Mo.)	450,000	oak-hickory forest	
40. Clark Nat'l Forest (Mo.)	902,000	oak-hickory forest	
41. Mingo Nat'l Wildlife Refuge (Mo.)		Mississippi River bottomland forest	

WILDLIFE SANCTUARIES IN THE CENTRAL UNITED STATES

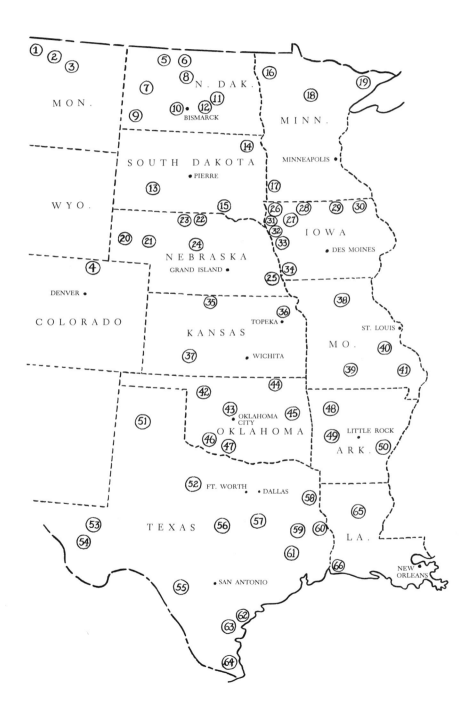

Central United States (cont'd)

	ACREAGE	VEGETATION	ANIMAL LIFE
42. Boiling Springs State Park (Okla.)	880	sand-dune vegetation	prairie chicken
43. Roman Nose State Park (Okla.)	560	mixed prairie, tallgrass prairie	bobcat, badger, beaver
44. Osage Hills State Park (Okla.)	1,000	oak-hickory woodland	bobcat, badger, coyote
45. Robbers Cave State Park (Okla.)	8,400	prairie & oak-pine forest	white-tailed deer, bobcat
46. Quartz Mtn. State Park (Okla.)	10,000	Wichita mountain oak-pine forest	coyote, badger
47. Wichita Mtns. Wildlife Refuge (Okla.)	61,000	mixed prairie, scrub-oak forest	bison, elk, white-tailed deer, turkey, antelope, coyote, bobcat
48. Ozark Nat'l Forest (Ark.)	1,045,000	oak-hickory-pine forest	
49. Ouachita Nat'l Forest (Ark.)	1,318,000	oak-hickory-pine forest	
50. White River Nat'l Wildlife Refuge (Ark.)	110,000	floodplain cypress forest	white-tailed deer, black bear, turkey
51. Palo Duro Canyon State Park (Texas)	15,000	oak-hickory & prairie	
52. Possum Kingdom State Park (Texas)	2,600	oak-hickory & prairie	
53. Balmorhea State Park (Texas)	50		
54. Davis Mtns. State Park (Texas)	1,800	oak-hickory-pine	antelope
55. Garner State Park (Texas)	640	oak-hickory	armadillo
56. Longhorn Cavern State Park (Texas)	700	oak forest with mesquite, yucca, red cedar	
57. Fort Parker State Park (Texas)	1,500		
58. Caddo Lake State Park (Texas)	500		
59. Davy Crockett Nat'l Forest (Texas)	161,000	coastal-plain forest	
60. Sabine Nat'l Forest (Texas)	183,000	bottomland forest, Sabine River	
61. Sam Houston Nat'l Forest (Texas)	158,000	coastal-plain forest	
62. Aransas Nat'l Wildlife Refuge (Texas)		Gulf coastal forest	waterfowl
63. Lake Corpus Christi State Park (Texas)	14,000	Gulf coastal forest	

Central United States (cont'd)

	ACREAGE	VEGETATION	ANIMAL LIFE
64. Santa Ana Wildlife Refuge (Texas)			
65. Kisatchie Nat'l Forest (La.)	494,000	coastal-plain forest	
66. Sabine Nat'l Wildlife Refuge (La.)		Gulf coastal forest	waterfowl

WILDLIFE SANCTUARIES IN THE ROCKY MOUNTAINS

	ACREAGE	VEGETATION	ANIMAL LIFE
1. Glacier Nat'l Park (Mont.)	1,008,000	alpine tundra, subalpine forest, montane & coastal forest transition	grizzly and black bears, white-tailed & mule deer, moose, elk, wolf, coyote, bobcat, mountain lion, bighorn sheep, mountain goat, lynx, wolverine
2. Cabinet Mtns. Wild Area (Mont.)	90,000	Cabinet & Kootenai Nat'l Forests; alpine tundra, subalpine forest	elk, black & grizzly bears, bighorn sheep, mountain goat, bobcat, wolf, mountain lion
3. Nat'l Bison Range (Mont.)	18,000	prairie, juniper woodland, montane forest	bison, mule & white-tailed deer, bighorn, elk, black bear, bobcat, mountain lion
4. Bob Marshall Wilderness Area (Mont.)	990,000	Flathead & Lewis & Clark Nat'l Forest; alpine tundra, subalpine & montane forest	mule & white-tailed deer, elk, moose, bighorn, mountain goat, wolf, grizzly & black bears, bobcat, lynx, mountain lion
5. Mission Mtns. Wild Area (Mont.)	75,000	Flathead Nat'l Forest; alpine tundra, subalpine forest	grizzly & black bears, mule & white-tailed deer, elk, wolf, bobcat, lynx, mountain lion
6. Anaconda-Pintlar Wilderness Area (Mont.)	145,000	Bitterroot & Deerlodge Nat'l Forest; alpine tundra, subalpine forest	moose, elk, black bear, mountain goat
7. Red Rock Lakes Migratory Waterfowl Refuge (Mont.)	32,000	lakes & marshes	trumpeter swan
8. Spanish Peaks Wild Area (Mont.)	50,000	Gallatin Nat'l Forest; alpine tundra, subalpine forest	white-tailed deer, bighorn sheep
9. Absaroka Wild Area (Mont.)	64,000	Gallatin Nat'l Forest; alpine tundra, subalpine forest	elk, moose, deer
10. Beartooth Wilderness Area (Mont.)	230,000	Custer Nat'l Forest; alpine tundra, subalpine forest	
11. Idaho Wilderness Area (Idaho)	1,232,000	Challis, Salmon, & Payette Nat'l Forests; subalpine & montane forest	white-tailed deer, elk, moose, mountain goat, bighorn sheep, black bear, wolf, coyote, bobcat, lynx
12. Sawtooth Wilderness Area (Idaho)	200,000	Challis, Boise, & Sawtooth Nat'l Forests; subalpine & montane forest	elk, mule deer, mountain goat

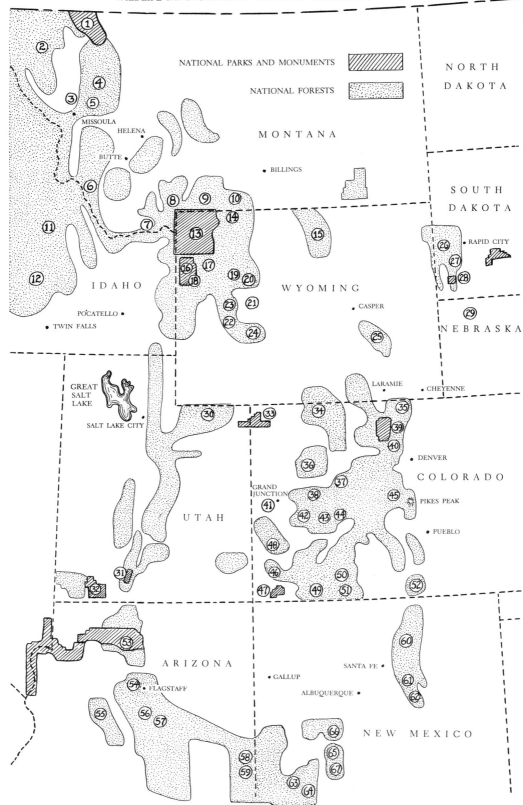

Rocky Mountains (cont'd)

	ACREAGE	VEGETATION	ANIMAL LIFE
13. Yellowstone Nat'l Park (Wyo.)	2,039,000	subalpine & montane forests; juniper woodland	grizzly & black bears, elk, antelope, moose, bighorn, bison, bobcat, lynx, coyote, trumpeter swan, pelican
14. North Absaroka Wilderness Area (Wyo.)	379,000	Shoshone Nat'l Forest; subalpine & montane forest	
15. Cloud Peak Wild Area (Wyo.)	94,000	Bighorn Nat'l Forest; subalpine forest	
16. Grand Teton Nat'l Park (Wyo.)	301,000	subalpine & montane forest	elk, moose, mule deer, grizzly & black bears, bighorn, wolverine, coyote, mountain lion, lynx
17. Teton Wilderness Area (Wyo.)	565,000	Teton Nat'l Forest; subalpine & montane forest	elk, moose, deer, bighorn, black bear
18. Nat'l Elk Refuge (Wyo.)		Grand Teton Nat'l Park	elk
19. South Absaroka Wilderness Area (Wyo.)	614,000	Shoshone Nat'l Forest; subalpine & montane forest	
20. Stratified Wilderness Area (Wyo.)	147,000	Shoshone Nat'l Forest; subalpine forest	
21. Wind River Mtns. Wilderness Area (Wyo.)	220,000	Shoshone Indian Reservation	
22. Bridger Wilderness Area (Wyo.)	383,000	Bridger Nat'l Forest; alpine tundra, subalpine forest	elk, deer, moose, bighorn, black bear
23. Glacier Wilderness Area (Wyo.)	177,000	subalpine & montane forest	
24. Popo Agie Wild Area (Wyo.)	70,000	subalpine & montane forest	mule deer, bighorn
25. Snowy Range Natural Area (Wyo.)	770	Medicine Bow Nat'l Forest; subalpine forest	
26. Upper Pine Creek Natural Area (S.D.)	1,200	Black Hill Nat'l Forest; montane forest	
27. Custer State Park (S.D.)	69,000	montane forest & chaparral	bison, elk, deer, bighorn, mountain goat, bobcat
28. Wind Cave Nat'l Park (S.D.)	27,000	montane forest & prairie	bison, elk, antelope, bobcat, mule deer, prairie dog, coyote
29. Chadron State Park (Nebr.)	640	conifer woodland	
30. High Uintas Wilderness Area (Utah)	243,000	alpine tundra, subalpine forest	mule deer, bighorn
31. Bryce Canyon Nat'l Park (Utah)	36,000	ponderosa forest, piñon-juniper woodland	mule deer, elk, black bear, coyote, prairie dog, mountain lion

324 / THE BIOMES OF NORTH AMERICA

Rocky Mountains (cont'd)

	ACREAGE	VEGETATION	ANIMAL LIFE
32. Zion Nat'l Park (Utah)	128,000	piñon-juniper woodland	mule deer, elk, bighorn, black bear, coyote, bobcat, mountain lion
33. Dinosaur Nat'l Monument (Colo.)	144,000	piñon-juniper woodland	mule deer, black bear, bobcat, bighorn, coyote, mountain lion
34. Mt. Zirkel-Dome Peak Wild Area (Colo.)	43,000	Routt Nat'l Forest; subalpine forest	
35. Rawah Wild Area (Colo.)	25,700	Roosevelt Nat'l Forest; subalpine forest	
36. Flat Tops Wilderness Area (Colo.)	117,800	White River Nat'l Forest; subalpine & montane forest	mule deer, elk
37. Gore Range-Eagle Nest Wild Area (Colo.)	61,200	White River and Arapaho Nat'l Forests; alpine tundra, subalpine forest	
38. Maroon-Snowmass Wild Area (Colo.)	64,600	White River Nat'l Forest; alpine tundra, subalpine forest	deer, elk, bighorn, black bear, wolf, mountain lion, lynx
39. Rocky Mtn. Nat'l Park (Colo.)	255,700	alpine, subalpine & montane forest	bighorn, elk, mule deer, black bear, coyote, mountain lion
40. Boulder Mtn. City Park (Colo.)	4,000	montane forest	
41. Colorado Nat'l Monument (Colo.)	18,000	piñon-juniper woodland	bison, elk, deer, coyote
42. Black Canyon Nat'l Monument (Colo.)	14,000	piñon, chaparral, Douglasfir forest	bighorn, mule deer, elk, coyote, black bear, bobcat
43. West Elk Wild Area (Colo.)	52,000	Gunnison Nat'l Forest; subalpine forest	
44. Gothic Natural Area (Colo.)	900	Gunnison Nat'l Forest; subalpine forest	deer, elk, wolf, mountain lion
45. Hurricane Canyon Natural Area (Colo.)	500	Pike Nat'l Forest; montane forest	
46. Narraguinnep Natural Area (Colo.)	2,800	San Juan Nat'l Forest; montane forest	
47. Mesa Verde Nat'l Park (Colo.)	51,000	piñon-juniper woodland, chaparral	mule deer, bighorn, bobcat, mountain lion, black bear, coyote, prairie dog
48. Uncompahgre Wild Area (Colo.)	69,000	Uncompahgre Nat'l Forest; subalpine forest	
49. Wilson Mtns. Wild Area (Colo.)	27,000	San Juan Nat'l Forest; alpine tundra, subalpine forest	

	ACREAGE	VEGETATION	ANIMAL LIFE
50. Upper Rio Grande Wild Area (Colo.)	56,600	Rio Grande Nat'l Forest; subalpine forest	
51. San Juan Wild Area (Colo.)	240,000	San Juan Nat'l Forest; subalpine forest	
52. La Garita-Sheep Mtn. Wild Area (Colo.)	38,000	Rio Grande, Gunnison, San Isabel Nat'l Forests	elk, bighorn
53. Grand Canyon Nat'l Park (Ariz.)	673,000	subalpine & montane forest, piñon-juniper woodland	mule deer, black bear, antelope, bighorn, coyote, bobcat, mountain lion
54. San Francisco Peaks Natural Area (Ariz.)	880	Coconino Nat'l Forest; subalpine forest	
55. Pine Mtn. Wild Area (Ariz.)	17,500	montane forest, oak-juniper woodland, chaparral	
56. Sycamore Canyon Wild Area (Ariz.)	47,000	Coconino, Kaibab, Prescott Nat'l Forests; montane forest & piñon-juniper woodland	
57. Oak Creek Canyon Natural Area (Ariz.)	940	Coconino Nat'l Forest; montane forest	
58. Mt. Baldy Wild Area (Ariz.)	7,400	Apache Nat'l Forest; subalpine & montane forest	
59. Blue Range Wilderness Area (Ariz.)	218,000	Apache & Crook Nat'l Forests; subalpine & montane forests	
60. San Pedro Parks Wild Area (N.M.)	41,000	Santa Fe Nat'l Forest; subalpine & montane forests	
61. Pecos Division Wilderness Area (N.M.)	137,000	Santa Fe Nat'l Forest; subalpine & montane forests	
62. Monument Canyon Natural Area (N.M.)	640	Santa Fe Nat'l Forest; montane forest	
63. Gila Wilderness Area (N.M.)	562,000	Gila Nat'l Forest; subalpine & montane forest, chaparral, prairie	deer, turkey, black bear, mountain lion
64. Black Range Wilderness Area (N.M.)	170,000	Gila Nat'l Forest; montane & subalpine forest	deer, black bear
65. Chupadora State Game Refuge (N.M.)	69,000	Cibola Nat'l Forest; piñon-juniper woodland, prairie	deer, antelope
66. Zuni Federal Game Refuge (N.M.)	46,000	Cibola Nat'l Forest; ponderosa pine forests, grassland	deer
67. Gallina-Bear Mtn. State Game Refuge (N.M.)	46,000	Cibola Nat'l Forest; ponderosa pine forest, piñon-juniper woodland	deer, black bear

WILDLIFE SANCTUARIES OF THE PACIFIC COAST STATES

	ACREAGE	VEGETATION	ANIMAL LIFE
1. Olympic Nat'l Park (Wash.)	888,000	coastal forest of spruce-hemlock-cedar	elk, black-tailed deer, black bear, mountain lion, mountain goat, bobcat, coyote, fisher
2. Quinault Natural Area (Wash.)	1,400	Olympic Nat'l Forest; coastal forest	black-tailed deer, elk, black bear, bobcat
3. Millersylvania State Park (Wash.)	830	coastal forest	black-tailed deer
4. Moran State Park (Wash.)	4,800	coastal forest	black-tailed deer
5. Larrabee State Park (Wash.)	1,300	coastal forest	black-tailed deer, black bear, bobcat, mountain lion
6. Deception Pass State Park (Wash.)	1,700	coastal forest	black-tailed deer, black bear, bobcat, mountain lion
7. North Cascade Wilderness Area (Wash.)	801,000	Chelan & Mt. Baker Nat'l Forests; coastal, montane, & subalpine forests	mountain goat, bighorn, black-tailed & mule deer, black & grizzly bears, lynx, coyote, wolf, mountain lion
8. North Fork Nooksack Natural Area (Wash.)	1,300	Mt. Baker Nat'l Forest; coastal forest	mountain goat, black-tailed deer, black bear, mountain lion
9. Lake Twenty-two Natural Area (Wash.)	800	Mt. Baker Nat'l Forest; coastal forest	black-tailed deer, bobcat, black bear, mountain lion
10. Long Creek Natural Area (Wash.)	640	"	black-tailed deer, black bear, bobcat, mountain lion
11. Meeks Table (Wash.)	90	Snoqualmie Nat'l Forest; ponderosa pine forest	
12. Mt. Rainier Nat'l Park (Wash.)	241,000	alpine tundra, subalpine forest	mountain goat, elk, black-tailed deer, mule deer, black bear, bobcat, coyote, mountain lion
13. Goat Rocks Wild Area (Wash.)	82,000	Pinchot & Snoqualmie Nat'l Forests; subalpine & montane forests	mountain goat, black-tailed deer, bobcat, black bear, mountain lion
14. Mt. Adams Wild Area (Wash.)	42,000	Pinchot Nat'l Forest; alpine tundra, subalpine & montane forests	black-tailed & mule deer, black bear, bobcat, mountain lion, marten
15. Cedar Flats Natural Area (Wash.)	700	Pinchot Nat'l Forest; coastal forest	black-tailed deer, elk, black bear, bobcat, mountain lion, coyote, marten
16. Beacon Rock State Park (Wash.)	3,000	coastal forest	black-tailed deer, black bear, bobcat, mountain lion

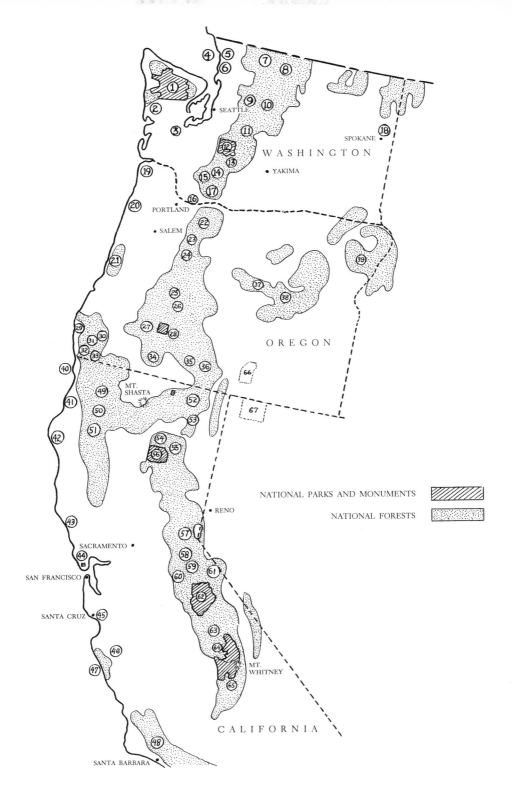

WILDLIFE SANCTUARIES ON THE PACIFIC COAST

Pacific Coast States (cont'd)

	ACREAGE	VEGETATION	ANIMAL LIFE
17. Wind River Natural Area (Wash.)	1,100	Pinchot Nat'l Forest; coastal-forest transition with montane forest	black-tailed deer, black bear, bobcat, mountain lion
18. Mt. Spokane State Park (Wash.)	2,300	coastal-forest transition with montane forest	white-tailed & mule deer, elk, black bear, bobcat, mountain lion, lynx
19. Saddle Mtn. State Park (Ore.)	3,000	seaside coastal forest	black-tailed deer, elk, black bear, bobcat, mountain lion
20. Cape Lookout State Park (Ore.)	1,400	coastal forest	black-tailed deer, black bear, bobcat, mountain lion, sea lion, seal
21. Neskowin Natural Area (Ore.)	700	Siuslaw Nat'l Forest; coastal forest	"
22. Mt. Hood Wild Area (Ore.)	14,000	Mt. Hood Nat'l Forest; subalpine forest	black-tailed deer, black bear, bobcat, mountain lion
23. Mt. Jefferson Wild Area (Ore.)	86,000	Deschutes, Mt. Hood, & Willamette Nat'l Forests; coast forest transition	black-tailed deer, black bear, bobcat, mountain lion
24. Three Sisters Wilderness Area (Ore.)	246,000	Deschutes & Willamette Nat'l Forests; subalpine & montane forests	black-tailed & mule deer, black bear, mountain lion, bobcat, wolf
25. Metolius Natural Area (Ore.)	1,400	Deschutes Nat'l Forest; subalpine & montane forests	mule deer, black bear
26. Pringle Falls Natural Area (Ore.)	1,100	Deschutes Nat'l Forest; montane forest	mule deer, black bear, mountain lion, bobcat
27. Abbott Creek Natural Area (Ore.)	2,600	Rogue River Nat'l Forest; montane forest	black-tailed deer, black bear, bobcat, mountain lion
28. Crater Lake Nat'l Park (Ore.)	160,000	montane forest	mule deer, black-tailed deer, elk, white-tailed deer, black bear, marten, red & gray foxes, bobcat, mountain lion, badger, marten
29. Cape Sebastian State Park (Ore.)	1,000	coastal forest	elk, black-tailed deer, black bear
30. Lobster Creek Natural Area (Ore.)	1,300	Siskiyou Nat'l Forest; Port Orford cedar, Douglasfir	
31. Coquille River Falls Natural Area (Ore.)	500	Siskiyou Nat'l Forest; coastal-forest transition	black-tailed deer, black bear, bobcat, mountain lion, coyote
32. Kalmiopsis Wild Area (Ore.)	78,000	Siskiyou Nat'l Forest; montane forest	black-tailed deer, black bear, elk, bobcat, mountain lion

Pacific Coast States (cont'd)

	ACREAGE	VEGETATION	ANIMAL LIFE
33. Port Orford Cedar Natural Area (Ore.)	1,100	Siskiyou Nat'l Forest; coastal-forest transition	black-tailed deer, bobcat, black bear, mountain lion
34. Mountain Lakes Wild Area (Ore.)	23,000	Rogue River Nat'l Forest; subalpine & montane forests	black-tailed & mule deer, black bear, bobcat, mountain lion
35. Gearhart Mtn. Wild Area (Ore.)	18,000	Fremont Nat'l Forest; subalpine & montane forests	mule deer, bobcat, black bear, mountain lion
36. Goodlow Mtn. Natural Area (Ore.)	1,200	Fremont Nat'l Forest; montane forest, juniper woodland	mule deer, black bear, bobcat
37. Ochoco Divide Natural Area (Ore.)	1,900	Ochoco Nat'l Forest; montane forest	mule deer, black bear, bobcat
38. Strawberry Mtn. Wild Area (Ore.)	34,000	Malheur Nat'l Forest; montane forest	black bear, bobcat, mule deer, elk
39. Eagle Cap Wilderness Area (Ore.)	223,000	Wallowa & Whitman Nat'l Forests; subalpine & montane forest	mule deer, black bear, bobcat, marten, mountain lion
40. Del Norte Coast Redwoods State Park (Cal.)	5,800	coast-redwood forest	mule deer
41. Prairie Creek Redwoods State Park (Cal.)	9,700	coast-redwood forest	mule deer
42. Humboldt Redwoods State Park (Cal.)	22,000	coast-redwood forest	mule deer
43. Armstrong Redwoods State Park (Cal.)	440	coast-redwood forest	mule deer
44. Muir Woods Nat'l Monument (Cal.)	485	coast-redwood forest & chaparral	mule deer, bobcat, gray fox
45. Big Basin Redwoods State Park (Cal.)	10,000	coast-redwood forest	mule deer, bobcat
46. Pinnacles Nat'l Monument (Cal.)	12,000	chaparral	mule deer
47. Pfeiffer-Big Sur State Park (Cal.)	780	coast-redwood forest	
48. San Rafael Wild Area (Cal.)	75,000	Los Padres Nat'l Forest; montane forest	
49. Marble Mtn. Wilderness Area (Cal.)	237,000	Klamath Nat'l Forest; montane forest, oak woodland	
50. Salmon-Trinity Alps Wilderness Area (Cal.)	280,000	Klamath, Shasta, Trinity Nat'l Forests; subalpine & montane forests, oak woodland	

Pacific Coast States (cont'd)

	ACREAGE	VEGETATION	ANIMAL LIFE
51. Middle Eel–Yolla Bolla Wilderness Area (Cal.)	143,000	Mendocino & Trinity Nat'l Forests; montane forest	
52. Devil's Garden Natural Area (Cal.)	16,000	Modoc Nat'l Forest; juniper woodland	
53. South Warner Wild Area (Cal.)	70,000	Modoc Nat'l Forest; montane forest, juniper woodland	
54. Thousand Lake Valley Wild Area (Cal.)	16,000	Lassen Nat'l Forest; montane forest	
55. Caribou Peak Wild Area (Cal.)	16,000	Lassen Nat'l Forest; montane forest	
56. Lassen Volcanic Nat'l Park (Cal.)	104,000	subalpine & montane forest	mule deer, black-tailed deer, black bear, bobcat, marten
57. Desolation Valley Wild Area (Cal.)	41,000	Eldorado Nat'l Forest; subalpine forest	
58. Harvey Hall Natural Area (Cal.)	4,200	Toiyabe Nat'l Forest; alpine tundra, subalpine forest	
59. Emigrant Basin Wild Area (Cal.)	98,000	Stanislaus Nat'l Forest; subalpine & montane forest	
60. Calaveras Big Trees State Park (Cal.)	5,300	montane forest; Sierra redwoods	
61. Hoover Wild Area (Cal.)	20,000	Toiyabe & Inyo Nat'l Forests; subalpine & montane forest	
62. Yosemite Nat'l Park (Cal.)	758,000	montane forest, Sierra redwoods	mule deer, black bear, marten, fisher, wolverine, red & gray foxes, badger
63. High Sierra Wilderness Area (Cal.)	826,000	Inyo, Sierra & Sequoia Nat'l Forests; subalpine & montane forest	
64. Kings Canyon Nat'l Park (Cal.)	453,000	montane forest	mule deer, bighorn, black bear, wolverine, mountain lion, fisher, marten
65. Sequoia Nat'l Park (Cal.)	385,000	montane forest, Sierra redwoods	mule deer, black bear, bighorn, mountain lion, coyote, fisher, marten, wolverine
66. Hart Mtn. Nat'l Antelope Refuge (Ore.)		montane forest, grassland	antelope
67. Charles Sheldon Antelope Range (Nev.)		montane forest, grassland	antelope

WILDLIFE SANCTUARIES IN THE SOUTHWESTERN UNITED STATES

	ACREAGE	VEGETATION	ANIMAL LIFE
1. Winnemucca Nat'l Wildlife Refuge (Nev.)		sagebrush desert	
2. Fallon Nat'l Wildlife Refuge (Nev.)		sagebrush desert	
3. Stillwater Nat'l Wildlife Refuge (Nev.)		sagebrush desert	
4. Humboldt Nat'l Forest (Nev.)	2,500,000	sagebrush desert, piñon-juniper	
5. Ruby Lake Nat'l Wildlife Refuge (Nev.)	35,000	sagebrush desert	grouse, mountain quail, mule deer
6. Toiyabe Nat'l Forest (Nev.)	2,480,000	sagebrush desert, piñon-juniper woodland	deer, mountain lion
7. Lehman Caves Nat'l Monument (Nev.)	640	sagebrush desert, piñon-pine woodland, ponderosa pine forest	
8. Cathedral Gorge State Park (Nev.)	1,500	sagebrush desert, piñon-juniper woodland	
9. Kershaw Canyon State Park (Nev.)	240	”	
10. Beaver Dam State Park (Nev.)	719	”	
11. Desert Game Range (Nev.)	2,022,000	”	bighorn, mule deer, elk
12. Valley of Fire State Park (Nev.)	8,000	”	
13. Locomotive Springs Nat'l Wildlife Refuge (Utah)		”	
14. Dinosaur Nat'l Monument (Utah)	46,000	sagebrush desert, piñon-juniper woodland	
15. Capitol Reef Nat'l Monument (Utah)	33,000	”	
16. Arches Nat'l Monument (Utah)	34,000	”	
17. Dixie State Park (Utah)	21,000	sagebrush desert, piñon-juniper woodland	
18. Natural Bridges Nat'l Monument (Utah)	2,600	sagebrush desert	
19. Death Valley Nat'l Monument (Cal.)	1,700,000	creosote bush & cactus-desert; piñon-juniper, montane forest	mule deer, bighorn, coyote, bobcat, kit fox, burro, desert rodents

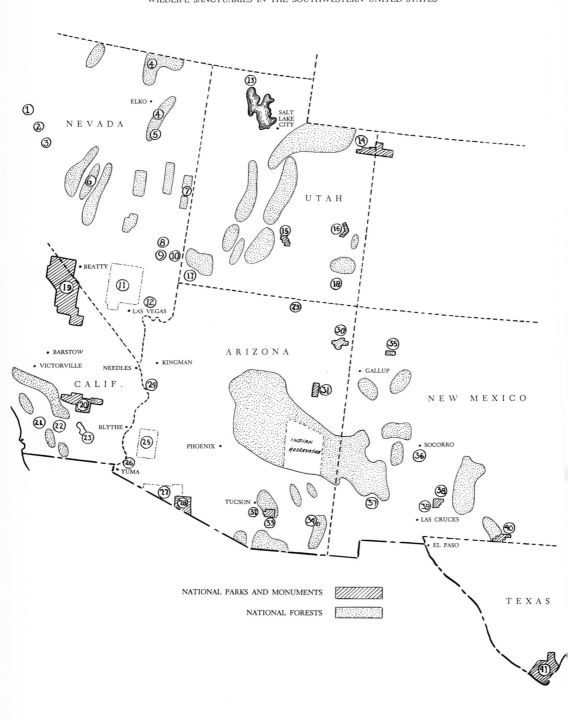

Southwestern United States (cont'd)

	ACREAGE	VEGETATION	ANIMAL LIFE
20. Joshua Tree Nat'l Monument (Cal.)	495,000	creosote bush & cactus desert; Joshua tree (yucca) forests	mule deer, bighorn, mountain lion, coyote, badger, kit fox, bobcat, desert rodents, desert reptiles
21. Salton Sea Nat'l Wildlife Refuge (Cal.)	32,400	creosote bush & cactus desert	waterfowl
22. Cuyamaca Rancho State Park (Cal.)	20,000	creosote bush & cactus desert	
23. Anza Desert State Park (Cal.)	251,000	creosote bush & cactus desert; California desert palm	desert rodents, desert reptiles
24. Havasu Lake Nat'l Wildlife Refuge (Ariz.)	37,000	on Colorado River, creosote desert vegetation	waterfowl
25. Kofa Game Range (Ariz.)	660,000	creosote bush & cactus desert; Arizona fan palm, saguaro cactus	desert bighorn, mule deer, peccary
26. Imperial Nat'l Wildlife Refuge (Ariz.)	47,000	on lower Colorado River	waterfowl
27. Cabeza Prieta Game Refuge (Ariz.)	860,000	creosote bush & cactus desert, saguaro	desert bighorn, mule deer, peccary, Gila monster
28. Organ Pipe Cactus Nat'l Monument (Ariz.)	328,000	creosote bush & cactus desert	mule and white-tailed deer, antelope, bighorn, coati, peccary, desert fox, coyote, bobcat, mountain lion
29. Navajo Nat'l Monument (Ariz.)	360	desert grassland, piñon-juniper woodland	
30. Canyon de Chelly Nat'l Monument (Ariz.)	83,000	"	
31. Petrified Forest Nat'l Monument (Ariz.)	85,000	creosote bush & cactus desert	
32. Desert Laboratory (Tucson, Ariz.)	1,900	creosote bush & cactus desert, saguaro	antelope, desert fox, desert reptiles & rodents
33. Saguaro Nat'l Monument (Ariz.)	60,900	creosote bush & cactus desert, saguaro	peccary, white-tailed deer, mule deer, coyote, mountain lion
34. Chiricahua Nat'l Monument (Ariz.)	10,000	creosote bush & cactus desert, piñon-juniper woodland	
35. Chaco Canyon Nat'l Monument (N.M.)	20,000	creosote bush & cactus desert, piñon-juniper woodland	
36. Bosque del Apache Nat'l Wildlife Refuge (N.M.)	56,000	Rio Grande valley; creosote bush & cactus desert	waterfowl

Southwestern United States (cont'd)

	ACREAGE	VEGETATION	ANIMAL LIFE
37. City of Rocks State Park (N.M.)	640	creosote bush & cactus desert	
38. White Sands Nat'l Monument (N.M.)	140,000	dune vegetation, yuccas	kit fox, coyote, desert reptiles & rodents
39. San Andres Nat'l Wildlife Area (N.M.)	57,000	piñon-juniper woodland	bighorn, mule deer, Gambel's quail, coyote, bobcat, golden eagle
40. Carlsbad Caverns Nat'l Park (N.M.)	45,800	creosote bush & cactus desert, piñon-juniper	deer

Index

Acacia, 58; sweet, 219
Air plants, 218, 225
Alder, 11, 13, 19, 21, 76, 88, 91, 114, 117, 118
Alfalfa, 62
Algae, blue-green, 38; green, 61–62
Alligator, 35, 95, 185, 213, 218; American, 232; blunt-nosed, 27
Amphibians, 13, 25, 40, 156, 158, 193, 293, 294; habitat, 43–44, 76, 106, 108
Anemone, 117
Angiosperm, 76, 293; classified, 16–17, 19; seed formation, 59
Anhinga, 29, 234
Animals, defined, 14
Annuals, desert, 267–268; tundra, 117
Anole, 25–26, 95, 230; green, 197–198
Antelope, 294; habitat, 52, 75; parasites of, 66; pronghorn, 32, 253–254
Antelope brush, 268
Anthrax, 65
Ants, 9
Aphids, 57, 65, 156
Appalachian Highlands, 96, 107, 108
Apple, 21
Arbor vitae, 17
Arctic-Alpine Life Zone, 104
Armadillo, 258
Ash, 88, 91; white, 178–179
Aspen, 106, 141, 176–177, 185
Associations, defined, 76
Aster, 249
Atlantic Coastal Plain, 98, 107, 109
Azalea, alpine, 118

Bacteria, 14, 33, 38, 55, 75, 81; nitrogen-fixing, 55–56, 62; parasite, 63–66
Badger, 258, 292
Balance of nature, 72
Baldpate, 130
Banana, 40
Basswood, 76
Bat, 50
Bayberry, 92
Bear, 33, 58, 123, 125, 127; big brown, 164; black, 33, 162–163, 230; grizzly, 33, 163–164; polar, 126–128, 293
Bearberry, 118
Beaver, 31, 58, 190, 192–193

Bee, 59
Beech, 8, 11, 13, 19, 52, 88, 172, 176, 177, 178, 180; American, 173
Beetle, 9, 57, 156; bark, 148; Engelmann spruce, 65; hickory bark, 186; Japanese, 67; leaf, 57, 186; wood-boring, 65, 81
Bighorn, desert, 286
Bilberry, 118
Biomass, defined, 81
Biome, defined, 75–76, 110, 112
Biosphere, defined, 72
Biotic community, 70–92; classification, 75–76; defined, 12, 70–72; energy flow, 77, 78–82; organization, 73–75; stability, 77, 83–92
Birch, 13, 19, 21, 39, 89, 91, 106, 117, 134, 172, 180; black, 173; dwarf, 141; gray, 176; paper, 176; 243; yellow, 173, 176
Birds, 27–30, 57, 66, 68, 294; conifer forest biome, 148–150, 156–158; deciduous forest biome, 186–189, 200–206; desert biome, 290; grassland biome, 251, 256–257; tropical biome 232–239; tundra biome 123, 127, 129–130. *See also* individual bird names

Cactus, 14, 23–24, 35, 47, 58, 106, 179, 218, 228–229, 241, 249, 263, 265; Arizona night-blooming, 48; barbed-wire, 228; barrel, 48, 275, 278; beavertail, 276; cane, 249, 275–276; cholla, 249, 275; Engelmann prickly pear, 276; hedgehog, 48, 278; jumping cholla 276–277; mistletoe, 228; nopal prickly pear, 276; organpipe, 275, 278–279; pincushion, 277–278; prickly pear, 95, 217, 228, 249, 262, 275–276; saguaro, 64, 275, 279–280, 294; staghorn cholla, 276; tree, 228
Canadian Life Zone, 107, 109
Carbon cycle, 53–55
Cardinal, 13, 94, 186
Caribou 32, 67, 81, 118, 127, 129, 130–131, 153, 156, 293; barren ground, 120–122; woodland, 123, 154
Carnivores, conifer biome, 155–164; deciduous forest biome, 193–210; desert biome, 287–292; energy transfer, 80;

336 / INDEX

grassland biome, 256–259; life pyramid, 82; niche, 75; tropic biome, 230–239; tundra biome, 125–129
Carolinian Life Zone, 106, 108
Cat family, 33, 68
Catalpa, 65, 177; northern, 177
Catbird, 186
Caterpillar, tent, 186
Catkin-trees, defined, 19; transfer of pollen, 59
Cat's-claw, 219, 271
Cattail, 11, 13, 45, 76, 91, 216
Cattle, 32, 65
Cedar, 17; incense, 143, 145; red, 17, 92, 175, 179; western red, 147; white, 17, 19
Central Lowlands, 98, 109
Central Lowlands forest 171–172, 177–179
Century plant, desert, 273–274
Cherry, 65, 92; sand, 91; wild, 21, 185
Chestnut, 19, 64–65, 67, 172; American, 174; horse, 21
Chickadee, 9, 13, 27, 149; black-capped, 202–203
Chicken, prairie, 251
Chipmunk, 11, 31, 35–37, 58, 81, 83, 157; eastern, 190–191; least, 151; yellow-pine, 151
Chuckwalla, 25–26, 282–283
Climate, 12, 13, 98–110; succession 85, 87. See also Physical environment
Climax community, defined, 85
Clover, 62
Coastal forest, 134, 145–147, 166–167, 172, 179–185
Commensalism, defined, 60
Condor, 27
Conifer forest biome, 112, 132–167; carnivores, 155–164; herbivores, 147–155; man, 164–167; physical environment, 132–133; plant life, 133–147
Conifers, 16–17, 19, 59, 74, 85, 87, 109, 293
Conservation, defined, 3
Coreopsis, desert, 268
Cottonwood, 59, 91, 177, 180; eastern, 177
Cougar, 33
Coyote, 33, 38, 75, 95, 106, 123, 259, 291, 294
Cranberry, mountain, 118
Crayfish, 158

Creosote bush, 47, 106, 265, 269
Cricket, 83, Mormon, 250
Crocodile, 25, 27, 95, 107, 213, 218, 231–233, 294; American, 27, 232
Crossbill, 149
Crow, 27
Cypress, 19, 106, 213; bald, 17, 45, 94, 180, 182–184, 215
Cypress forest community, 217

Dandelion, 75
Darwin, Charles, 2
Deciduous, defined, 12
Deciduous forest biome, 112, 169–211, 293; carnivores, 193–210; herbivores, 185–193; man, 210–211; physical environment, 170–185
Deer family, 32, 58, 67, 68–69, 134, 156, 157, 210, 218, 293; Arizona white-tailed, 286; desert mule, 286; Key, 229; mule, 123, 154; white-tailed, 153, 190, 229
Desert biome, 112, 294; carnivores, 287–292; herbivores, 280–287; man, 292–295; physical environment, 265–267; plant life, 267–280; regions, 263–265
Dicotyledon, 19, 21, 23
Diurnal animals, defined, 50
Dog family, 33, 68; prairie, 31, 35, 52, 75, 95, 108, 253, 294
Dogberry, 8
Dogwood, 8, 13, 21, 94, 106, 172, 175, 177, 179, 180, 185, 215
Dormancy, 41
Duck, 29
Dust storm, 261–262

Eagle, 30, 156; bald, 157; golden, 157
Ecology, defined, 1–2; history of, 2–3; man's role in, 4–5; name, 2–3; research in, 3–4
Ecosystem, 72–73; defined, 72; energy flow, 78–82; food chain, 80–81; life pyramid, 81–82
Egret, 95, 213; cattle, 60; common, 235; snowy, 235–236; white, 29, 183–184
Elk, 32, 41, 106, 154
Elm, 65, 92, 95, 178
Entomologist, 11
Epiphytes, defined, 43; tropical biome, 218, 224–228
Estivation, defined, 42
Everglades community, 216

Evergreen trees, 39, 47, 66, 106, 108, 109, 118, 293; broad-leaved, 218–221

Falcon, gyrfalcon, 123, 128–129; peregrine, 129; prairie, 290
Fawn, 9
Fern, 8, 9, 13, 14, 43, 50, 52, 88, 134, 218; epiphytic, 218; marsh, 216; royal, 216
Fiddleneck, 268
Fig, strangler, 215, 218, 227–228
Finch, 149; American goldfinch, 203; goldfinch, 27, 57, 251
Fir, 12, 17, 104, 134, 136, 293; balsam, 137, 176; Douglas, 52, 143
Fisher, 160
Flamingo, 107; American, 238–239
Flea, 66; rat, 67
Flicker, gilded, 290; yellow-shafted, 204
Florida fish-poison tree, 219
Fly, 66; deer, 156; flesh, 67; green-headed, 156; horse, 156; louse, 66; warble, 67
Food chain, 80–81, 113, 185–186, 239, 293; birds, 201; carnivores, 125; effect of fallout, 131; lemming, 120; mink, 160; parasite, 81; polar bear, 128; rock ptarmigan, 123; saprophyte, 81
Fox, 33, 58, 65, 157, 293; kit, 258–259, 292; red, 9, 33, 208–209, 210; white, 126
Frog, 25, 38, 44, 83, 91, 94, 186, 193; green treefrog, 194; leopard, 193–194; wood, 9, 125
Fungus, 33, 63–65, 75, 81; bracket, 33; shelf, 33

Gallinule, purple, 235
Gila monster, 25–26, 288–289
Glasswort, 51
Goat, mountain, 32, 52, 123, 124–125, 129
Goldenrod, 92, 117, 249
Goldfields, 268
Goose, 29; cackling, 130; lesser Canada, 130; white-fronted, 130
Gopher, 58; northern pocket, 252; plains pocket, 252; western pocket, 285–286
Grama, blue, 248
Grape, possum, 218
Grass, 14, 22, 38, 51, 59, 74, 81, 88, 89, 92, 95, 104, 108, 114, 115, 117, 118, 213, 216, 241, 245, 246, 293; big bluestem, 246; buffalo, 244, 248–249, 262; desert, 106; June, 244, 247; Kentucky, 246; little bluestem, 246–247; needlegrass, 244; sawgrass, 216; western wheatgrass, 248
Grass-sedge community, 117–118
Grasshopper, 57, 281
Grassland biome, 112, 294; carnivores, 256–259; classified, 241–243; herbivores, 250–256; man, 260–262; physical environment, 243–244, 260; plant life, 244–250
Grosbeak, 149
Grouse, 9; ruffed, 13, 28, 149, 186
Grub, 9, 57
Gull, 27
Gymnosperm, 16–17, 59

Haeckel, Ernst, 2–3
Hammock forest community, 218
Hare, 31, 75, 118, 120, 128, 129; arctic, 120–121; varying, 152
Hawk, 30, 68, 81, 83, 156; goshawk, 157; ferruginous, 256–257; osprey, 30; red-shouldered, 206; red-tailed, 205, 256, 290; sparrow, 290
Hawthorne, 21
Hedgehog, 275
Hellbender, 195
Hemlock, 8, 11, 17, 52, 88, 176, 178, 180; eastern, 172, 175; western, 145, 147
Hen, heath, 188
Herbaceous plants, defined, 14, 59
Herbivores, 17, 56–58, 85, 92; conifer biome, 147–155; deciduous forest biome, 185–193; desert biome, 280–287; energy transfer, 80–82; grassland biome, 250–256; niche, 74–75; tropical biome, 229–230; tundra biome, 120–125
Heron, 29; blue, 11; great blue, 205; white, 236–237
Hibernation, 41
Hickory, 8, 13, 19, 52, 85, 88, 92, 95, 106, 172, 177, 178, 179, 180; shagbark, 173, 176
Holly, 12, 106, 172, 180, 215; American, 176
Host, defined, 63
Hyacinth, water, 45, 213, 215
Hydrophytes, defined, 44–45

338 / INDEX

Ibis, 29, 95; white, 237; wood, 236
Intermontane Plateau, 98, 109
Insect, 57, 66–67, 186, 250, 281; gall, 186; scale, 65–66
Insectivore, 30
Invertebrate, defined, 24
Ironwood, 8, 13
Isohyet, defined, 101, 103, 109–110
Isotherm, defined, 100, 103

Jaguar, 33, 292
Jay, 27, 149; blue, 186; Canada, 150; Rocky Mountain, 150; Steller's, 150
Juniper, 17, 52, 91, 96, 106, 243

Kangaroo, 30
Katydid, 83
Killifish, 38
Kingfisher, belted, 204
Kite, Everglade, 239; swallowtail, 239

Labrador tea, 118
Ladybug, 156
Larch, 17, 45, 51
Lark, 251; eastern meadowlark, 203–204
Laurel, 8, 172
Laurentian Plateau, 98
Legume, 62
Lemming, 31, 120, 126, 128, 129, 157, 293
Lemon vine, 228–229
Lichen, 14, 24, 39, 43, 61, 74, 81, 88, 104, 114, 115, 117, 131, 134, 293; reindeer, 217
Lichen-moss community, 117–118
Life pyramid, 81–82
Life zone, 103–107
Lignum vitae, 219
Lily, pond, 91, 216
Limpkin, 235
Linden, 91
Lion, mountain, 33, 68, 123, 125, 161–162, 230, 290, 293
Lizard, 25, 41–42, 95, 106; crested, 281–282; fence, 25–26, 94, 197, 230, 256; horned, 25–26, 49, 287; ocellated sand, 288; spiny, 288; whiptail, 288
Locust, 57; black, 177, 178; Rocky Mountain, 250
Locust tree borer, 186
Longspur, Lapland, 27
Loon, 29, 91

Louisianian Life Zone, 106, 108
Louse, 66, 67
Lower Sonoran Life Zone, 106–107, 109
Lupine, 117
Lynx, 33, 68, 125, 160; Canada, 160–161

Magnolia, 12, 21, 95, 106, 185; southern, 180–182
Mahogany, West Indies, 219
Mallard, 130
Mammal, 30–33, 49, 58, 66, 67–69; grassland biome, 257–259; clawed, 32–33; desert hoofed, 286–287; gnawing, 30–31; hoofed, 31–32; primitive, 30
Mammoth, 88
Man, conifer forest biome, 164–167; deciduous forest biome, 210–211; desert biome, 292–295; tundra biome, 130–131; grassland biome, 260–262; tropical biome, 240
Manatee, 38, 107, 213, 218, 230
Manchineel, 218, 219–220
Mangrove, 95, 107, 213, 215, 294; black, 221; red, 220–221
Mangrove forest community, 217–218
Maples, 8, 12, 13, 19, 21, 35–36, 39, 59, 76, 88, 91, 94, 95, 106, 176; red, 13, 45, 92, 174, 176, 177, 215; silver, 174; sugar, 47, 88, 174, 176, 177, 178, 180
Marigold, desert, 268
Marmot, yellow-bellied, 151–152
Marsupial, 30
Marten, 33, 124, 160
Massasauga, 256
Mastodon, 88
Mayflower, Canada, 8
Meadowlark, 27
Mesophyte, defined, 44–47
Mesquite, 58, 74, 106, 179, 241, 243, 250, 262; honey, 271; screwbean, 271
Metabolic water, defined, 49
Migration, 40–41
Mildew, 34
Millipede, 185
Mink, 33, 65, 67, 158–159
Mistletoe, 63
Mockingbird, 13, 94, 186
Mold, 14, 34
Mole, 30
Mollusk, 229
Monocot, defined, 19, 21–22
Montane forest, 134, 141–145, 166

Moose, 32, 58, 75, 104, 134, 153, 154–155, 156
Mormon tea, 269
Mosquito, 156
Moss, 8–9, 13, 14, 24, 43, 50, 52, 104, 115, 117, 134, 218; Spanish, 43, 60, 88, 94, 180, 224–225, 294
Moth, 57; gypsy, 67, 186; tussock, 148, 186
Mouse, 9, 31, 120, 157; deer, 251; desert deer, 284; harvest, 229, 251; jumping, 31
Mudpuppy, 195
Mushroom, 14, 34
Muskeg, 134
Muskox, 32, 118, 122–123, 127, 129, 130, 293
Muskrat, 11, 91, 158, 190, 191
Mutualism, 60–61
Mycelium, defined, 64

Natural selection, law of, 69
Needle and thread, 247–248
Nematode, 65
Newt, 25, red-spotted, 195
Niche, defined, 73
Nitrogen cycle, 53–56, 62
Nocturnal animals, 50
Northern Appalachian forest 171, 176–177
Nuthatch, white-breasted, 201–202

Oak, 8, 12, 13, 19, 52, 74, 76, 81, 88, 91, 92, 94, 95, 106, 172, 176, 177, 178, 179, 180, 215; black, 19, 172–173; eastern live, 181; live, 19, 95, 96, 180, 218; red, 172, 176; scrub, 217; white, 19, 172–173, 176
Ocelot, 33, 292
Ocotillo, 269
Oldsquaw, 130
Opossum, 13, 30, 94, 207–208, 293
Optimum temperature, defined, 38
Orchid, 19, 43, 107, 213, 215, 218; cowhorn, 226; greenfly, 226; spread-eagle, 226–227; white butterfly, 227
Oriole, 27
Osprey, 68, 157
Owl, 30, 50, 68, 83, 156; barred, 206; boreal, 157–158; burrowing, 257; elf, 290; great gray, 158; horned, 9; great horned, 158, 290; screech, 206; short-eared, 257; snowy, 127, 128

Pacific Coast Ranges, 96, 98, 109
Palm, 19, 21, 95, 107, 108, 221–224, 294; cabbage, 95, 216, 218, 221–222; California fan, 271–272; date, 40; dove, 219; royal, 215, 222–223
Palmetto, 94, 106, 180, 216; saw, 217, 222
Paloverde, common, 271; yellow, 271
Panther, 33
Papaya, 95
Parakeet, Carolina, 189
Parasite, 33, 63–67
Peccary, 286–287
Pelican, brown, 239
Pepper vine, 218
Perennial, herbaceous, 85, 117, 294; woody, 117
Permafrost, 116; defined, 39
Photosynthesis, 14, 35–36, 38, 47, 49, 58, 61–62, 63, 119, 277
Physical environment, conifer forest biome, 132–133; deciduous forest biome, 170–185; desert biome, 265–267; grassland biome, 243–244, 260; tropical biome, 214–215; tundra biome, 115–117
Physiographic succession, 88
Pickerelweed, 216
Pigeon, domestic, 67; passenger, 67, 188–189
Pika, 123–124, 129
Pincushion, 275
Pine, 13, 16–17, 76, 88, 91, 94, 95, 106, 134, 180, 213, 215, 293; bristlecone, 138, 140–141; jack, 178; loblolly, 180; lodgepole, 17, 138; longleaf, 180; piñon, 52, 106; pitch, 138; ponderosa, 17, 52, 95, 96, 143, 148, 179; red, 178; sand, 217; slash, 180–181, 216; southern, 180; sugar, 52; white, 17, 67, 137–138, 176, 178
Pine forest community, 216–217
Pintail, 130
Pitcher plant, 51
Plague, bubonic, 65, 67
Plant bug, 57
Plant life, conifer forest biome, 133–147; grassland biome, 244–250; tropical biome, 215–229; tundra biome, 117–119. *See also* Physical environment
Plant reproduction, 59
Plants, green, 14–24
Plum, 21; coco, 219

Poisonwood, 218, 219
Poplar, 19, 88, 134; aspen, 243; tulip, 106
Poppy, 249
Porcupine, 9, 13, 31, 152, 190
Prairie, mixed, 241, 247–250; true, 241, 246–247
Precipitation, 101, 170–171, 214, 243, 265–266. *See also* Physical environment
Predator, 67–69
Primrose, desert, 268
Protozoa, 62, 66; coccidia, 66; trypanosome, 66
Psoralea, silverleaf, 248–249
Ptarmigan, 128, 129, 293; rock, 28, 123
Puma, 33
Pussy willow, 19
Pyramid of life, 129, 155, 193, 259, 293, 294

Quail, bobwhite, 28, 187; mountain, 149

Rabbit, 30–31, 65, 134, 157, 158, 210, 229, 293; antelope jackrabbit, 286; cottontail, 9, 190–191, 229; desert, 286; desert cottontail, 286; eastern cottontail, 253; jackrabbit, 49, 109, 253, 286, 290
Raccoon, 9, 76, 83, 95, 207, 210, 218, 293
Ragweed, 92
Raspberry, 134
Rat, 31; bannertail kangaroo, 284–285; cotton, 229; Florida water, 229; kangaroo, 31, 49, 294; pack, 31, 49, 285; rice, 31, 229
Ratten, 218
Redbud, 94, 172, 175, 177, 179, 180, 215
Redstart, American, 203
Redwood, 52, 293; Coast, 17, 147; Sierra, 17, 19, 145
Reindeer, 67, 122, 130
Remora, 60
Reptile, 13, 25–27, 40, 41–42, 48–49, 106, 108, 193, 196–200, 287–290, 293, 294; conifer forest biome, 156; desert biome, 281–284; grassland biome, 256; tropical biome, 230–232
Rhizome, defined, 246
Rhododendron, 172
Roadrunner, 42, 290
Robin, 94
Rocky Mountains, 98

Rodents, 30–31, 58, 65, 85, 106, 118, 134, 158, 229, 284–286, 293, 294
Rubber vine, 218
Rush, 91, 216

Sagebrush, 47, 96, 106, 109, 250, 263; purple, 268
Salamander, 9, 25, 44, 193; dusky, 195; spotted, 195
Saltbush, 268–269; desert, 51
Saprophyte, 33–34
Saxifrage, 117
Scamp, 130
Scorpionweed, 268
Scoter, American, 130
Scrub-birch community, 117–118
Sea grape, 219
Seal, 128
Sedge, 11, 91, 117
Seed, dispersal, 59–60; production, 116; plants, 14–16
Sheep, bighorn, 32, 123, 125, 129; Dall, 123, 125
Shrew, 30
Shrub, nonsucculent, 268–269
Sierra Nevada, 109
Skink, 25
Skunk, 33, 83; striped, 207; spotted, 292
Slope, 51–52
Smoke tree, 271
Snake, 25, 26, 106, 186; black, 9; copperhead, 26, 200, 256; coral, 26, 200, 231; cottonmouth, 26, 183, 200, 231; Great Plains rat, 256; garter, 198, 256; gopher, 289; indigo, 231; kingsnake (scarlet), 231, (western), 289; leaf-nosed, 289; milk, 198; pit viper, 26; racer, 198, 256, 289; rattlesnake, 26, 106, 198, 200, 231, 256, 289; sidewinder, 289–290; water moccasin, 231
Soapweed, 250
Society, defined, 76
Soil, acid, 51; alkaline, 51; effect on flora, 13; desert, 266–267. *See also* Physical environment
Solar energy, 78, 80
Southern Appalachian forest, 171, 172–176
Sparrow, 57, 251; English, 75; white-throated, 94, 149
Spoonbill, roseate, 237–238
Spore plants, defined, 14; simple spore plants, 84–85

Sprengel, Christian, 2
Spring peeper, 194
Spruce, 12, 16–17, 35, 81, 88, 94, 95, 104, 107, 134, 136, 293; black, 117; blue, 52; Engelmann, 95, 141, 148; red, 136–137, 148, 176; white, 117, 136, 243
Squirrel, 31, 58, 65, 67, 76, 83, 91, 95, 157, 293; Douglas, 150; fox, 13, 190; golden-mantled ground, 151; gray, 190, 191, 229; ground, 120, 290; red, 9, 13, 94, 150, 190; striped ground, 257–258; tassel-eared, 150–151; white-tailed antelope, 285
Stick-tights, 60
Stolon, defined, 246
Subalpine forest, 134, 138–141, 166
Succulents, desert, 273–280
Swan, 130
Sweetgum, 12, 185
Switchgrass, 246
Sycamore, 21, 177, 180
Symbiosis, defined, 60

Tamarack, 17
Tamarind, wild, 219
Teal, 130
Temperature, 38–39, 40–42, 98, 100; desert biome, 266; grassland biome, 244; tropical biome, 214. *See also* Physical environment
Temperature belts, 100–101
Temperature, optimum, 40
Termites, 62
Thistle, 249; Russian, 262
Thrush, 83, 91, 149; hermit, 149; Swainson's, 149–150
Tidy-tip, 268
Tiller, defined, 246
Titmouse, tufted, 203
Toad, 25, 83, 193; Hudson Bay, 125
Tolerance, defined, 49
Tortoise, 26–27; desert, 283; gopher, 26, 27, 197
Towhee, 9
Transition Life Zone, 104, 106, 107, 108, 109
Transpiration, defined, 39, 43, 47–48
Tropical biome, 112, 293; carnivore, 230–239; herbivore, 229–230; man, 240; physical environment, 214–215; plant life, 215–229

Tropical Life Zone, 107, 108
Tularemia, 65
Tuliptree, 12, 21, 76, 172, 174, 177, 180
Tundra biome, 110, 113–131, 293; arctic, 114–115, 117, 118–119; carnivores, 125–130; classification, 114–115; herbivores, 120–125; man, 130–131; physical environment, 115–117; plant life, 117–119
Turkey, 28; wild, 186, 187, 229
Turtle, 25, 26–27, 94, 186, 213; box, 27; common spotted, 26–27; eastern box, 197; mud, 230–231; painted, 27, 196; pond, 230–231; snapping, 196–197; wood, 27
Typhus, 67

Ungulates, 31–32, 58, 66, 67
Upper Sonoran Life Zone, 106, 108–109

Vanilla vine, 227
Verbena, 268
Vertebrates, cold-blooded, 24–27; defined, 24–25; warm-blooded, 25, 40, 104
Viburnum, 8, 92
Vireo, 27; red-eyed, 186
Vole, 31, 120; prairie, 251
Vulture, 27; black, 206; turkey, 206

Walnut, 19, 172, 177, 178; black, 173
Walrus, 128
Wapiti, 123
Warbler, 27, 149
Wasp, 59
Water, 13, 42–49
Water birds, 28–30
Weasel, 11, 32–33, 68, 81, 123, 124, 125; short-tailed, 158
Whale, 128
Whip-poor-will, 83
Wildlife sanctuaries, Alaska and Canada, 297–300; Central U.S., 316–320; Mideastern U.S., 304–308; Northeastern U.S., 301–303; Pacific Coast, 326–330; Rocky Mountains, 321–325; Southeastern U.S., 309–315; Southwestern U.S., 331–334
Willows, 19, 21, 88, 91, 106, 114, 117, 118, 134, 180, 185; arctic, 141; dune, 91
Witch hazel, 65
Wolf, 33, 68, 81, 293; gray, 33, 122, 126–127

Wolverine, 33, 68, 104, 123, 125, 293
Woodchuck, 31, 94, 190, 191–192
Woodpecker, 27, 81, 156; arctic three-toed, 148; downy, 9; Gila, 290; ivory-billed, 202
Worms, 9, 66, 185, 229; cutworm, 57; flatworm, 66; hookworm, 66; liver fluke, 66; pinworm, 66; roundworm, 66; screwworm, 67; segmented worms, 66; spruce budworm, 65; tapeworm, 66
Wren, cactus, 27

Xerophyte, 44, 47, 48, 263, 294

Yucca, 19, 21–22, 51, 60–61, 96, 241, 274; Joshua tree, 274